The media's watching Vault!
Here's a sampling of our coverage.

"For those hoping to climb the ladder of success, [Vault's] insights are priceless."
– *Money magazine*

"The best place on the web to prepare for a job search."
– *Fortune*

"[Vault guides] make for excellent starting points for job hunters and should be purchased by academic libraries for their career sections [and] university career centers."
– *Library Journal*

"The granddaddy of worker sites."
– *U.S. News and World Report*

"A killer app."
– *New York Times*

One of Forbes' 33 "Favorite Sites"
– *Forbes*

"To get the unvarnished scoop, check out Vault."
– *Smart Money Magazine*

"Vault has a wealth of information about major employers and job-searching strategies as well as comments from workers about their experiences at specific companies."
– *The Washington Post*

"A key reference for those who want to know what it takes to get hired by a law firm and what to expect once they get there."
– *New York Law Journal*

"Vault [provides] the skinny on working conditions at all kinds of companies from current and former employees."
– *USA Today*

VAULT
> the most trusted name in career information™

VAULT CAREER GUIDE TO THE
ENERGY
INDUSTRY

VAULT CAREER GUIDE TO THE
ENERGY INDUSTRY

LAURA WALKER CHUNG
AND THE STAFF OF VAULT

ACKNOWLEDGMENTS

We are extremely grateful to Vault's entire staff for all their help in the editorial, production and marketing processes. Vault also would like to acknowledge the support of our investors, clients, employees, family, and friends. Thank you!

Wondering what it's like to work at a specific employer?

Read what EMPLOYEES have to say about:

- Workplace culture
- Compensation
- Hours
- Diversity
- Hiring process

Read employer surveys on THOUSANDS of top employers.

Table of Contents

Visit Vault at **www.vault.com** for insider company profiles, expert advice,
career message boards, expert resume reviews, the Vault Job Board and more.

VAULT CAREER LIBRARY   ix

Chapter 8: Breaking Down the Jobs — 111

APPENDIX — 139

Visit Vault at **www.vault.com** for insider company profiles, expert advice,
career message boards, expert resume reviews, the Vault Job Board and more.

VAULT CAREER LIBRARY   xi

Use the Internet's
MOST TARGETED
job search tools.

Vault Job Board

Target your search by industry, function, and experience level, and find the job openings that you want.

VaultMatch Resume Database

Vault takes match-making to the next level: post your resume and customize your search by industry, function, experience and more. We'll match job listings with your interests and criteria and e-mail them directly to your inbox.

Introduction

The energy sector constitutes some 5% of the U.S. economy, generating revenues in excess of $1.5 trillion and employing more than one million people. We are all too well aware of how fundamental the energy sector is to the economy when a small change in world oil prices causes interest rates to jump and the stock market to shudder. Energy companies face and address questions of crucial importance to the economy and to all of our individual lives; their actions affect our government foreign policy, the quality of our environment, our ability to travel and work, the cost of nearly everything we purchase, and the health of our families. During the boom years of the 1990s, exceptional growth and technological change in the energy sector showed it to be an unusually influential, even glamorous place to be. The energy world offers enormous opportunity for the business job seeker.

However, the energy sector is notoriously complicated to understand and difficult to penetrate. Energy science and economics is not usually taught in business programs, and insiders say it takes years to truly get one's arms around the complexities of these topics. Having a reasonably intelligent conversation with a prospective employer requires a broad range of technical knowledge and comfort with the issues of the day, so this is not an industry where you can expected to get hired based on smarts alone. This book lays out not only a big-picture description of the opportunities, but also the details of what you need to know on a technical level to put yourself in the running. It will provide you the knowledge you need to determine if energy is the right fit for you, identify prospective employers, and position yourself for success as a job applicant.

Visit Vault at **www.vault.com** for insider company profiles, expert advice, career message boards, expert resume reviews, the Vault Job Board and more.

VAULT CAREER LIBRARY

1

THE SCOOP

Industry Overview

CHAPTER 1

What is the Energy Sector?

The energy sector produces, converts and distributes fuels to produce heat, light and propulsion. Oil, natural gas, and coal are burned to make heat and electricity. Wind, flowing water, and sunlight are converted into electricity. Oil is refined to propel cars, planes, and industrial machines. And to achieve these things, the companies who are producing, transporting, converting and distributing these energy sources are supported by a variety of service firms, investors, equipment providers, and government regulators. (See Figure 1.1)

There is a great divide in the energy sector between the oil and gas "side" and the electricity "side," each of which accounts for about half of the business jobs across the sector. "Oil and gas" refers to the exploration for and extraction and processing of oil and natural gas. In contrast, the electric power business revolves around converting fuel to electricity in power plants and distributing that electricity to consumers. The economics of the two fields, and the regulations that govern them, are quite distinct. Generally, people make their energy careers in one camp or the other, without too much crossover. Natural gas is one arena that bridges the oil & gas versus electricity divide – it is extracted from the earth together with oil, and is also a primary fuel for generating electricity.

When people refer to the "energy sector," they can actually mean any of the following: electric power, oil & gas, or both together. This guide takes a broad view of the industry, covering upstream (exploration), midstream (refining) and downstream (distribution and sales) oil and gas activities, electric power generation and transmission, equipment manufacturing, regulatory oversight, and lending to, investing in, and advising companies involved in the sector.

Just how big is the industry that comprises all those diverse activities? Companies in the energy sector take in nearly $1 trillion in revenue annually, out of the $17 trillion earned by all U.S. businesses. Energy-related businesses employ about 2.5 million people, or 2% of the U.S. workforce – far more than banking, high tech or telecommunications. Energy companies as a whole employ a high percentage of production workers (the people who drive local utility repair trucks, laborers on oil rigs, and gas station attendants), compared to other industries; of the 2.5 million energy jobs in the U.S., about 90% of them are blue-collar jobs or technical positions. The

Visit Vault at www.vault.com for insider company profiles, expert advice, career message boards, expert resume reviews, the Vault Job Board and more.

VAULT CAREER LIBRARY   5

subject of this guidebook is the one-quarter-million energy-related business jobs out there: the business analysts, finance associates, marketing managers, economic modelers, and operations consultants, to name a few roles.

Energy sector positions capture about 2% of new MBA graduates, an amount roughly proportional to the industry's size. In contrast, the investment banking and investment management sectors together capture 40% of graduates, and consulting absorbs another 20%. Even the significantly smaller high tech industry takes on 3 times the number of new MBAs as does the energy sector. What this means for you as a job seeker is that the energy sector is not as dominated by people with graduate business degrees as some other popular arenas. There is plenty of opportunity for smart, well-trained college graduates to rise through the ranks without necessarily going back to school.

Sector	US employees in managerial, business or financial positions
Pharmaceuticals and biotechnology	50,000
Telecommunications	140,000
High technology	200,000
Banking and investment management	250,000
Energy	250,000
Consulting	500,000
Entire economy	11,500,000

Figure 1.1: The Energy Value Chain

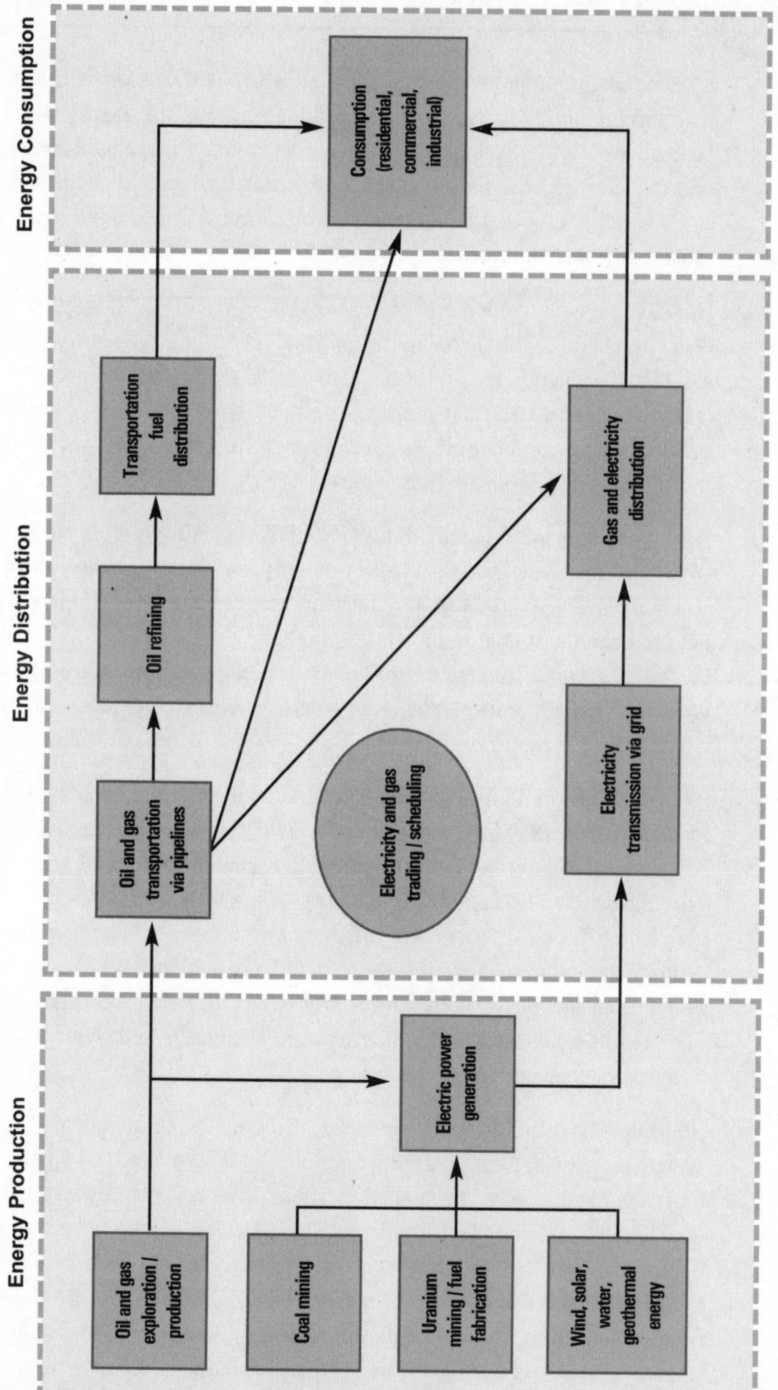

Visit Vault at **www.vault.com** for insider company profiles, expert advice,
career message boards, expert resume reviews, the Vault Job Board and more.

VAULT CAREER LIBRARY

7

Industry History

It's shocking to think that the Middle East, which is such a perpetual focus of U.S. foreign policy has not always been the center of the energy world. In fact, oil was only discovered in the region in the 1950s. Only some two decades later, demand for oil had skyrocketed in tandem with the new supply, and a cartel had been formed that controlled world prices tightly enough to cause a severe economic crisis in the U.S.

How long ago did contemporary methods of generating heat, light, and work come into being? Compared to the information technology sector, you might say that the energy sector is an old industry. However, compared to the majority of industries that make up our economy — banking, publishing, construction, manufacturing, to name a few — the energy sector as we know it is a recent development. (See Figure 1.2)

After spending hundreds of thousands of years burning wood to heat our caves and then our houses, wood eventually became scarce and expensive, and humans discovered the slow-burning heat of coal. The Chinese figured out the benefits of coal in the 100s A.D., followed by the Europeans during the Middle Ages. Access to coal quickly became a European geopolitical issue so volatile that it sparked bitter conflicts between Germany and France, who battled for centuries over Alsace and the coal-rich Saar Valley.

At first, people burned coal to heat the air directly, but in the 1800s coal-fired radiant hot water heating systems proliferated, relegating the sooty mess coal created to the basement. In the early 20th century, a natural gas pipeline system started to be laid down, allowing homes to burn a far cleaner fuel to heat either the air or water for radiators. Europe, with limited natural gas deposits, turned to oil-fired radiant hot water heating once oil became readily available in the early 1900s. After World War II, when electricity became more reliable and far cheaper, houses were eventually built with all-electric heating systems, particularly in Europe.

Not only was burning wood our earliest heating source, it was also our first source of artificial light. Lighting became a little more constant when animal fat-based candles were developed in about 3000 B.C., followed closely by liquid animal fat and plant oil lamps (think Aladdin's lamp, or the oil lamps lighting ancient Biblical temples). Around the world, evenings were lit by flickering flames until the turn of the 19th century, when coal gas lighting was first introduced to affluent homes and public sidewalks. In the mid-1800s, when petroleum oil was discovered and drilled in the U.S., it was refined into kerosene to produce a higher-quality oil lamp for the masses.

The revolution in lighting was, of course, electric light. Electricity was first produced in the mid-1800s, but was only used to power industrial machinery at first. When practical incandescent light bulbs were commercialized in the 1880s, indoor electric lighting quickly spread around the world.

Where did the electricity to power the light bulb come from? The first electric power plant was a coal-fired steam turbine generator built in 1880. However, after the success of the first hydroelectric power plant at Niagara Falls in 1895, reliable, clean hydropower provided most of the electricity in the U.S. in the first half of the 20th century. As electricity demand grew, many coal plants were built, and some countries developed geothermal power infrastructure (using steam from deep in the earth to drive turbines). Natural gas-fueled power plants first entered the mix after World War II, when the pipeline infrastructure was robust enough to provide a constant fuel source. In 1957, the first nuclear reactor started operation. In recent decades, commercial solar power, windpower and fuel cells reappeared, after having been first developed experimentally in the 19th century.

In contrast to the late 19th century revolution in indoor lighting, the watershed years in harnessing energy to do work occurred far earlier. After relying for thousands of years on waterwheels, windmills and sails to turn gears and propel objects (and before that on animals and our own brute strength), humankind saw the development of the steam engine at the turn of the 18th century. These external combustion engines used wood or coal to boil water into steam, which turned gears that ran factories, drove trains, and ultimately spawned the Industrial Revolution. Ironically, factories operating such relatively sophisticated machinery in 1750 would have been lit only by smoky oil lamps and heated by sooty coal stoves.

With the advent of electric generation in the mid-1800s came electric and battery-powered motors (in which the batteries were charged by electricity) that drove industrial machinery of all types. Even before electricity was applied to lighting, it was used to drive the very first cars – a design that effectively fell by the wayside until 1997, when Toyota introduced a commercial hybrid electric car. The commercial development of oil-fired internal combustion engines in the late 1800s allowed cars to go faster and farther, initiating our society's seemingly insatiable appetite for petroleum.

By the early 1900s, most people in the industrial world had access to clean, radiant indoor heating and constant, bright indoor incandescent electric lighting. They drove gasoline-powered cars with internal combustion engines, and had run their factories with powerful steam turbines for nearly two centuries. Oil and natural gas wells across the U.S. and Europe were

Visit Vault at **www.vault.com** for insider company profiles, expert advice, career message boards, expert resume reviews, the Vault Job Board and more.

V/\ULT CAREER LIBRARY

9

pumping in earnest, and major new deposits were soon to be discovered in the Middle East.

Since then, the major methods by which we produce heat, light, and work have not materially changed. The history of the energy sector since the mid-1900s has been a story of technological advancements in efficiency and environmental impact reduction, as the industry transformed from a low-tech, heavy-manufacturing identity to one that is fast-paced, cutting-edge, and very high-tech.

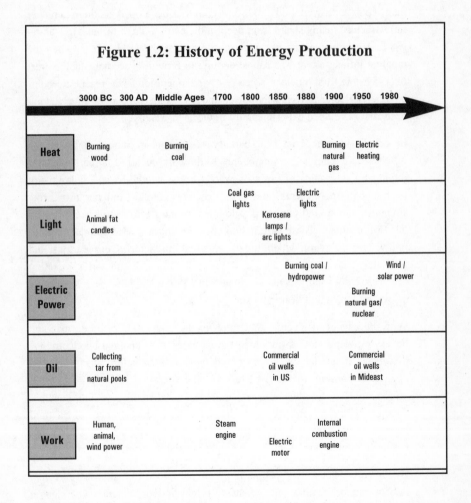

Figure 1.2: History of Energy Production

Technology Frontiers

People in the energy sector are passionate about the high-tech nature of their industry. Although the basic process of gathering fuels and burning them to produce electricity, light, heat and work is fairly set, how efficient that process is and how much pollution it generates are the subjects of some pretty hard-core science and revolutionary changes.

In order to be a compelling job candidate, it helps to be knowledgeable and passionate about new technologies appearing on the horizon. Below are just a handful of examples of the cutting-edge research and development happening in different parts of the energy world:

- One company has developed a very low-tech alchemy to turn pollution into fuel: letting algae "eat" NOx (nitrogen oxide) and CO2 (carbon dioxide) emissions from fossil plants. Algae thrive on these abundant feedstocks in power plant exhaust streams – pollution is reduced, the algae grows, and can later be dried and burned as fuel.

- An important emerging power plant pollution-control technology involves burning fuel without using a flame. The recently-commercialized Xonon combustor turns natural gas into energy by bringing it into contact with a catalyst. In flameless combustion, no NOx is formed at all, thus preventing smog and acid rain.

- High-temperature superconducting transmission lines are currently in testing. Traditional copper wires cause enough resistance to lose some 10% of the electricity they carry. Superconducting niobium-titanium alloy wires cooled by a tiny liquid nitrogen core eliminate line losses and thus effectively increase our electricity supply.

- We have used the principal of piezoelectricity for decades to generate electricity from motion. Certain types of miniature crystals spontaneously generate a high-voltage current when moved – portable gas grill lighters that don't use a flint work in this way. Now companies are looking into more advanced applications, like using piezoelectric generators embedded in the sole of a soldier's boot to power battlefield equipment.

- Did you know there's a gasoline-powered car with a fuel efficiency of 10,000 MPG? Such impressive efficiency comes from using a tiny, one-cylinder engine with little internal resistance, thin hard tires and a super-light car body to reduce road friction, a bubble-shaped aerodynamic design and a lightweight child driver on a smooth and level indoor track. While these extreme design elements may not be practical for commercial

Visit Vault at **www.vault.com** for insider company profiles, expert advice,
career message boards, expert resume reviews, the Vault Job Board and more.

VAULT CAREER LIBRARY   1 1

vehicles, fuel efficiency R&D (research & development) is an active and promising space.

- Data transmission over electrical wires is a little-known, older technology that is finding exciting new applications. The current in an electrical wire can also carry data – street lights have been remotely controlled this way for decades, and in the past few years home automation via the wires in the wall has proliferated (did you know that you can turn your dishwasher on from your computer at work?). Companies have recently started marketing modems that send and receive more complex data over power lines: Internet access, voice, and video.

- Battery technology is one of the real stumbling blocks of technological innovation these days – how much does it help you to have a supercomputer the size of a pad of paper if it dies after being unplugged for two hours? Companies are bringing better rechargeable lithium ion batteries to market, and actively developing laptops and cell phones powered by fuel cells with mini onboard tanks of hydrogen or methanol fuel.

- If you think wireless electricity is impossible, think again! We will soon be seeing desk surfaces and other furniture manufactured with embedded electrical chips – when you put a portable device down on the surface, the chip activates and recharges your laptop, phone, television, blender, razor, vacuum cleaner…or whatever. Less realistically, people have also thought about wireless electricity in the form of "space solar power," in which huge solar-paneled satellites would collect energy from the sun and beam it to earth in the form of radio frequencies, which would then be converted into electricity.

- Bringing the fabled hydrogen economy to reality requires an inexpensive source of free hydrogen. Development-stage hydrogen production techniques include harnessing the sun to release hydrogen from pure sugar, and using high-temperature catalysis (rather than energy-intensive electrolysis) to split water into H_2 and O_2.

Energy Concepts Overview

To prepare for a job search in energy, you need to learn a lot about how the technologies and markets work, as well as adopt your prospective employers' working vocabulary. Using terms like "secondary recovery," "heat rate" or "stack" confidently and correctly in context help you seem like a candidate who could step right into the job. Study up on the whole industry, regardless of which aspect of it you are targeting – for example, even in an oil & gas interview, a deep understanding of the electrical grid can provide you valuable context and give you an edge. This chapter will give you an overview of the most important concepts in energy.

Fuel Sources

Electricity can be generated from a wide variety of fuels: natural gas, coal, uranium (nuclear power), oil, wind, water (hydroelectric power, tidal power), solar radiation (photovoltaic power, solar thermal power), volcanic heat (geothermal power), plant material (biomass power), or hydrogen (fuel cells). The mix of fuel sources for electric generation varies greatly around the world, reflecting each country's natural resource base as well as its politics. (See Figure 1.3)

The United States gets about half of its electricity from coal, and most of the remainder from nuclear and natural gas. Right next door, Canada has chosen to leverage more of its rivers for hydropower, while exporting its gas to the U.S. and leaving most of its coal deposits intact. Norway, also with plentiful water resources and no indigenous coal (but plenty of offshore oil), gets all of its power from hydroelectric stations.

Australia relies overwhelmingly on coal, and has never ventured into nuclear power or developed renewables. In striking contrast, France made the explicit decision in the late 1970s to focus the vast majority of its electricity sector on nuclear power, despite the ready availability of coal in Europe. Not only does France see nuclear power as a clean alternative to fossil fuels, but it has managed to control the costs of nuclear plants and consequently to enjoy cheap electricity.

The Philippines is a world leader in renewable energy use, due to its exploitation of Pacific Rim volcanic activity for geothermal power; as

Visit Vault at **www.vault.com** for insider company profiles, expert advice, career message boards, expert resume reviews, the Vault Job Board and more.

VAULT CAREER LIBRARY   13

electricity demand has skyrocketed in recent years, though, coal was developed to meet needs, and is now also a primary fuel. Denmark also has one of the most highly developed renewable sectors in the world, focused predominantly on windpower.

Sitting on top of the largest oil reserves in the world, Saudi Arabia uses oil to generate most of its power. The abundance of cheap oil makes development of the country's other main natural resource – the sun – unattractive. Iran, in contrast, sits atop enormous gas reserves, which it has exploited to supply the majority of its electricity.

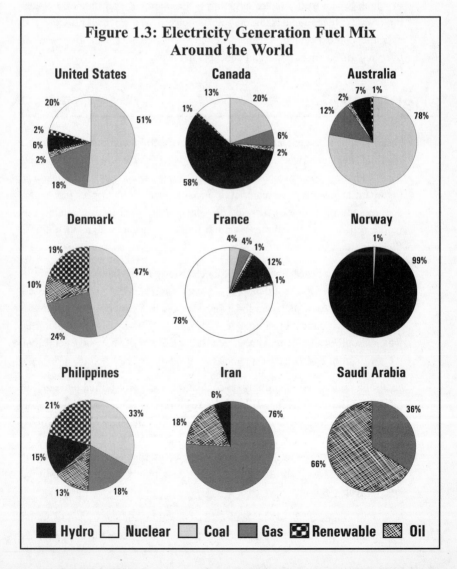

Figure 1.3: Electricity Generation Fuel Mix Around the World

Converting Fuel to Electricity

Power plants generate electricity through the magic of electromagnetism: spinning a strong magnet inside coils of conductive wire to produce a current. The magnet, sitting at the end of a shaft on a cylindrical turbine, spins at a rate of 30 times per second when water, air, steam or hot gas pushes against the turbine's many protruding blades. Power plants use three types of turbines:

1. **Steam turbines.** Nuclear, solar thermal, geothermal, biomass and most fossil fuel plants use steam (heated by combusting oil, gas, or coal, or by fissioning uranium atoms) to turn their turbines.

2. **Combustion turbines.** Simple-cycle combustion turbines use hot combustion gasses to spin. In gas-fired combined cycle plants, hot gas is used to spin one turbine, and then the combustion heat boils water into steam to turn a second turbine – using the gas twice dramatically increases the overall efficiency of these plants.

3. **Direct drive turbines.** Hydroelectric plants use the energy of falling water to turn turbines, just like a waterwheel at a 19th century riverside mill. Similarly, windpower plants harness the wind with blades 100 feet long to turn their magnetic rotors.

There are, however, two types of electricity generation technologies that don't use a turbine at all:

- **Fuel cells** use a chemical reaction to produce electricity. Natural gas or hydrogen molecules release their electrons in the presence of a catalyst metal, and then reformulate into water molecules while also releasing heat. The flow of electrons (rather than the rotation of a magnet) creates an electric current, and the hot water output can be used for heating applications. No combustion or turbine is involved, so fuel cells are quiet, motionless electric generators.

- **Photovoltaic cells** also generate electricity through a chemical reaction: photons of sunlight knock electrons in a silicon semiconductor away from their nuclei; the resulting flow of electrons is an electric current that is collected onto copper electrical wires.

Heat rates measure the efficiency of fossil fuel power plants in terms of British thermal units (BTUs, a measure of energy content) of fuel input required to produce 1 kilowatt-hour (kWh) of electrical output. The heat rate degrades over time as an ageing plant loses its edge; it also varies based on

Visit Vault at **www.vault.com** for insider company profiles, expert advice, career message boards, expert resume reviews, the Vault Job Board and more.

VAULT CAREER LIBRARY   **15**

the ambient air density (temperature) and the operating output level of the turbine. "New and clean" heat rates for the most advanced combined-cycle plants are currently 6500 BTU/kWh, and design improvements continue to slowly improve efficiency. Gas steam plants are substantially less efficient, with heat rates around 8,000; coal-fired and simple-cycle plant heat rates can exceed 12,000. Fuel cells usually have heat rates around 7,000 – 8,000 BTU/kWh.

> Electricity is most typically generated by burning fuel to boil water, and using the resulting steam to spin a large magnet, creating an electric current.

The Grid

Electricity is the most complex of all commodities, due to four peculiar characteristics:

1. Electricity is subject to virtually inelastic demand, which leads to extreme price volatility and market manipulability. (In other words, consumers don't vary the amount of electricity they demand based on its price; a factory owner, for example, will turn lights and equipment on in the morning no matter what. As any Microeconomics 101 textbook accurately predicts, if demand stays constant when prices go up, prices can spike up dramatically.)

2. Electricity cannot be stored easily, and thus must be produced the instant it is consumed.

3. Electricity cannot be routed to a destination. It flows along the path of least resistance, no matter how we might try to control it.

4. Transmission losses of electricity are substantial, which means that it must be produced relatively close to where it will be consumed.

Demand for electricity varies cyclically across the course of a day (peaking during the daytime), a week (peaking on workdays), and a year (peaking in summer due to air conditioning use, and sometimes in winter due to electric heating). Peak annual demand (a hot summer weekday afternoon) can be double that of minimum demand (pre-dawn hours on a mild spring day).

> The grid transmits electricity output from 14,000 baseload, intermediate load and peaking plants.

What results from this demand variation is a heterogenous fleet of some 14,000 power plants, with 980,000 Megawatts (MW) of generating capacity, dotting the United States. The majority of these are "baseload" plants that essentially run all the time, except for maintenance downtime. Then, there are intermediate load or load-following plants that are designed to cycle on and off as their region's demand fluctuates up and down over the course of a week. Finally, peaking plants – designed with a range of startup times, from 30 minutes, to 10 minutes, to instantaneous (spinning reserves) – provide the electricity to meet changes in demand within a day and within an hour, as each of us contributes to vacillating system load by nonchalantly flipping lights and appliances on and off. Why aren't all plants designed to start up quickly and thus be able to respond to demand fluctuations? As it turns out, the physics of power generation dictates a direct tradeoff between ramp-up time and efficiency; quick-start peaker plants are extremely inefficient (which means extremely expensive to run as well as highly polluting), and many of them thankfully only need to run a few hours every year.

Storing electricity, in theory, would preclude the need for peakers. Instead, baseload plants could generate a constant electricity output, store the excess when demand is low, and release that stored energy when demand is high. Unfortunately, none of our available electricity storage alternatives are cost effective:

- Pumped storage facilities use cheap nighttime output to pump water up a hill, storing potential energy to be released in the form of hydroelectric power during the day. However, energy expended to pump the water often offsets savings from avoiding peak-hour generation.

- Electricity can charge a battery, storing chemical energy. Batteries, however, are highly inefficient.

- Flywheels can store electricity in the form of kinetic energy for very short periods of time only, and with substantial losses.

- Finally, electricity can split water into hydrogen gas, which can be stored for later use in a fuel cell or combustion engine. For now, though, the process is expensive, and we have little infrastructure to store, transport, or combust hydrogen gas.

Visit Vault at **www.vault.com** for insider company profiles, expert advice, career message boards, expert resume reviews, the Vault Job Board and more.

VAULT CAREER LIBRARY

17

The electricity generated by power plants is fed into "the grid," a 160,000-mile network of insulated copper wires. In fact, due to idiosyncrasies of history, our U.S./Canadian transmission system was laid out as three semi-independent grids that have very few connections to one another: the Eastern Interconnect (everything east of the Rocky Mountain foothills), Western Interconnect, and Texas. Each grid is a virtual island of highly interdependent power lines – when one line fails, it can affect the entire interconnect.

> The U.S. transmission system is considered to be inadequate and increasingly unreliable.

Power failures have become more common in the past few years, due to a severe lack of investment in maintaining and improving the system. In August 2003, the worst blackout in American history left most of the eastern U.S. and Canada without electricity for several hours, apparently because some tree limbs in Ohio brought down one transmission line, which then tripped other lines in rapid succession across a span of one thousand miles. The successful operation of this fragile transmission system is fundamental to our economy and our individual lives; in recognition of this, the National Academy of Engineering named the power grid as the single greatest technological advance of the 20th century (followed in second place by the car).

Upgrading our inadequate power grid is one of the hot issues in the energy world today. How can we ensure electrical reliability? The only choices are to increase supply of transmission capacity, or decrease demands on the system. For example:

• **Increase supply**

  – Motivate companies to invest in new transmission lines, through regulatory mandates, government subsidies, or private "merchant" investment opportunities

  – Upgrade existing copper wires to superconducting cables that can handle larger energy flows

• **Decrease demand**

  – Develop distributed generation (generation near the point of consumption) on a widespread basis

– Install emerging "smart grid" technologies that manage energy flows to minimize looping, losses, and system trips

Because the grid is not a unified, continuous system across the country, the market price of electricity is not constant across the country either. Bottlenecks in the system result in a couple dozen or so (depending on how strictly you define them) regional markets; the unique demand patterns and supply levels within each market determine the market price. Forecasting these regional electricity prices is the subject of much debate, study, and concern. Unless you can accurately forecast long-term prices, you cannot make good decisions about whether and when to build new plants or engage in long-term contracts.

Most participants in the electricity marketplace today use sophisticated software programs to forecast prices. These programs operate by creating a "stack" of available supply for each day of the year, and compare it to demand for that day using a simulation of economic dispatch to generate a theoretical market-clearing price. The software user then generates a variety of possible scenarios by layering in complicating factors such as plants shut down for maintenance, changing fuel commodity prices, spiking demand due to weather changes, and uneconomic bidding by generators. Ultimately, interpreting the model output into a baseline forecast for use as an input to decisions throughout the company is a subjective and artful process.

Energy Prices

Oil is a commodity, and thus pre-tax prices for the same grade are fairly similar around the world. Those prices have hovered around $20 per barrel (in today's dollars) since World War II, though spiking as high as $80 in the late 1970s, and close to $50 as of this writing. Prices refer to a standardized quality of oil at a specific location; oils with different weight, sulfur or viscosity characteristics, or different delivery points, are all priced in reference to a few benchmark prices. West Texas Intermediate is the benchmark for the Western hemisphere, Brent blend is the reference price for Europe and Africa, and the OPEC basket is used in much of the rest of the world. For example, a barrel of heavy, "sour" (high sulfur) oil from Venezuela might sell for WTI minus $4.

Gasoline, an oil derivative, sells for about $1 per gallon wholesale, and currently about $2 per gallon at the U.S. retail pump, once taxes, transportation, and retailer operating costs are added in. In Europe, retail

Visit Vault at **www.vault.com** for insider company profiles, expert advice, career message boards, expert resume reviews, the Vault Job Board and more.

V/\ULT CAREER LIBRARY   **19**

gasoline is about four times as expensive as in the U.S., due to much higher taxes.

In contrast to oil, natural gas is not easily transported overseas, so pricing is specific to local markets. Over land, natural gas is transported via pressurized pipelines. For overseas shipping, natural gas is compressed into a liquid form (liquefied natural gas, LNG) so that more BTUs can fit on a single ship. Natural gas transportation is more expensive than oil transport, and thus a more substantial component of the total delivered price. Gas in the U.S. is referenced off of Louisiana's Henry Hub location, where prices have ranged from about $2 – $5 per million BTUs (MMBtu) over the past several years. Transportation of the gas to California could easily cost another few dollars per MMBtu.

Oil and gas drilling operations start up and shut down at various fields as the commodity prices fluctuate. When oil or gas prices are relatively high, extracting oil from tar sands or gas from shale is economically viable. However, when prices fall, those operations cannot cover their costs and are idled.

Electricity pricing is far more complex than that of oil and natural gas. Wholesale prices are specific to each region on the grid, and vary across the day, week, season and year, averaging somewhere on the order of $30 per MWh. Retail electricity prices are regulated, and change only when your local utility receives permission from the state Public Utility Commission. Consumers in the U.S. pay as little as 5 cents per kWh ($50/MWh) in the Pacific Northwest, where cheap hydroelectricity dominates, to as much as 17 cents per kWh in Hawaii, where a lack of natural resources forces the island to rely on inefficient and expensive oil-fired generation systems. U.S. retail electricity prices are among the lowest in the world.

Nuclear Power

Nuclear power is generated using the same steam turbines as in fossil fuel (oil or natural gas) plants. However, in a nuclear reactor, water is heated into steam by fissioning uranium atoms, rather than burning hydrocarbons.

The world's supply of uranium ore is mined primarily in Canada (40% of world supply) and Australia (20%). Because ore actually contains only 0.3% to 12% uranium, it is processed into a solid cake of 85% "natural uranium" (U3O8), which is a standardized commodity that sells for about $20/kg. Natural uranium is then converted into uranium hexafluoride ("hex," UF6), in which format it can be enriched. Naturally-occurring uranium contains primarily stable U-238, and less than 1% of the fissile U-235 isotope; enrichment increases the U-235 percentage to between three and five percent. After enrichment, the hex is converted into solid pellets of uranium oxide (UO2), which are sealed inside helium-filled metal tubes and shipped to nuclear power plants.

Fuel is used for about four years in the reactor until it is spent. Then, it is either reprocessed for reuse, or stored as waste in concrete-lined ponds, pending final storage in a repository. Reprocessing is a controversial practice by which spent uranium fuel is transformed back into fissile uranium oxide (96% of the original waste), fissile plutonium oxide (1%), and "final waste" (3%). Reprocessing is appealing because it greatly extends the life of uranium fuel, resulting is less uranium mining and thus preventing the associated aquifer contamination. So why is the process controversial?

1. **Proliferation.** Plutonium is the fuel for nuclear weapons, and just a few kilos are required to arm a bomb. While the plutonium generated by reprocessing is not weapons-grade, it could be converted into such form if it fell into ill-intentioned hands. Reprocessed plutonium iseither used by governments for weapons, permanently stored in carefully-monitored government facilities, or mixed with uranium to make a mixed-oxide "MOX" reactor fuel. The plutonium in MOX is rendered useless for weapons purposes after reuse.

2. **Health impacts.** Fuel is reprocessed by boiling it in nitric acid, generating significant amounts of sludge that is far more radioactive than the original nuclear waste. Existing reprocessing facilities release massive amounts of radioactive caesium and technetium into the ocean. As a result, concentrations of radionuclides several times higher than maximum recommended amounts can be found in lobsters, mussels, and seaweeds near the power plants. The amount of radioactivity in liquid wastes stored at reprocessing plants can be up to 100 times the amount released by the Chernobyl disaster; a serious accident at one of these plants would result in more than one million new cancer deaths. Even under normal operations, cancer clusters have developed near many of the plants.

Currently, only France, Russia, India, and Japan reprocess waste, collectively treating about 10% of annual worldwide spent fuel. North Korea may also be reprocessing waste explicitly for weapons development; and the fear surrounding Iran's plans to develop a nuclear power industry is that it will do the same. The UK is in the process of shutting down its reprocessing plant,

Visit Vault at **www.vault.com** for insider company profiles, expert advice, career message boards, expert resume reviews, the Vault Job Board and more.

VAULT CAREER LIBRARY   **21**

and the U.S. shut down its facilities in the 1970s due to concerns about plutonium proliferation, worker safety, and waste generation. However, the U.S. and Russia are now developing a MOX fuel fabrication plant in South Carolina as part of a non-proliferation strategy.

In the 1950s, nuclear power was famously anticipated to be "too cheap to meter" (the then head of the Atomic Energy Commission used the phrase in a speech extolling the promise of the new U.S. nuclear power program). Yet today it is one of the more costly sources of electricity in the U.S. system. Given the affordability of uranium, and the small quantities needed to fuel a reactor, one might reasonably expect the power output to be cheap. However, the cost of managing spent fuel (which has a radioactive half-life of 100 years), the cost of safety monitoring, and the much-higher-than-expected capital costs of building each custom plant have driven the actual cost of power up substantially.

Research and development of next-generation nuclear power plants is ongoing in the U.S. and many other countries. Future nuclear reactors may use naturally-abundant thorium as a fuel, or use faster neutron movement to utilize normally non-fissile U-238. On a more experimental level, research into controlled fusion as an energy production means also continues (fusion is how the sun produces energy, combining hydrogen to form helium).

Depending on whom you talk to, you will hear that the nuclear power industry is essentially "dead," or that it is merely stalled and may see renewed activity in the next few years:

Nuclear power has fallen out of favor...	...but still offers benefits to consider.
• No new orders for nuclear plant construction have been placed in the U.S. since 1973 • Most of Europe is in the process of phasing out nuclear power: banning new plants and decommissioning existing ones • Nuclear plants provide baseload power, but only because they take a long time to ramp on and off, not because they produce cheap power • The problem of long-term storage problem of radioactive waste has not been solved: Yucca Mountain in Nevada has been designated as the US final repository; but concerns over groundwater contamination remain.	• Nuclear energy generates no direct air pollution or greenhouse gases (though the fuel mining and fabrication require energy expenditures that do have emissions themselves). If the U.S. were to replace all of its nuclear plants with gas plants, CO_2 emissions would increase by more than 1% • Meeting targets for reducing greenhouse gas emissions may be difficult via energy efficiency and renewable power development alone • Until a new "surprise" technology appears, nuclear power can solve the problem of meeting exponentially-rising demand for electricity in the face of finite oil and gas supplies

Nuclear power has fallen out of favor...	...but still offers benefits to consider.
• Nuclear plants are horrific accidents waiting to happen (e.g. Chernobyl), and potential terrorist targets. Japan, for example, had to temporarily close down 17 plants for safety lapses in 2003.	• With respect to cost, France for one was able to set up its nuclear sector to produce cheap electricity by adopting just one type of nuclear technology for all of its plants • Canada and Japan still find reasons to be actively pro-nuclear, and many east Asian countries are interested in developing nuclear power sectors

Nuclear is properly pronounced 'NEW-klee-ur.' Never say 'NEW-kyoo-lur' in an interview!

Finding Oil

Locating underground petroleum deposits (usually referred to more simply as "oil") is primarily the work of engineers and geologists. Most commonly, the presence of oil is detected through seismology: an air gun is shot into the water, or heavy plates are slammed onto the ground, sending shock waves down into the ground or ocean floor. The speed and angle of the bounced-back shock waves reveal characteristics of the material they passed through on their journey. In addition, high-tech "sniffers" and instruments detecting small magnetic and gravitational field variations are also used to identify possible oil deposits. Ultimately, though, the only way to know for sure if oil exists in a given area is to drill an exploratory well, typically called a "wildcat." Despite all the best geological analysis, in the U.S. today only 1 in 40 wildcats succeeds in finding oil.

Oil reservoirs are not pools of liquid oil, but simply areas of solid rock with billions of droplets of oil trapped inside the rock's pores. If one drills a hole into such rock (i.e., a well), pressure in the surrounding rock will cause oil to seep out of the pores into the empty hole. The oil is then removed in three phases:

- **Primary production.** About 25% of the oil trapped in a given rock formation can be extracted by simply pumping out the oil that seeps into the well.

- **Secondary recovery.** Another 15% of the trapped oil can often be extracted by injecting water into the formation through a second well, increasing pressure on the deposits and squeezing more oil from the rock. In many cases, methane gas (known as "natural gas" or simply "gas") is

Visit Vault at **www.vault.com** for insider company profiles, expert advice, career message boards, expert resume reviews, the Vault Job Board and more.

VAULT CAREER LIBRARY 23

collocated with oil deposits. This gas can be extracted and sold, flared off as waste if the drilling company has no cost-effective means of getting it to market, or reinjected into the oil well in place of water for secondary recovery purposes.

- **Enhanced recovery.** An additional 30% of the existing oil may be accessible through expensive tertiary extraction methods. Pumping steam, carbon dioxide (naturally occurring in the oil deposit, or sequestered from an industrial waste stream), or gas-producing microbes into an injection well further increases the pressure on the oil deposit, squeezing more oil droplets through the rock's pores into the well. Surfactants and polymers can also be injected to break oil droplets away from the rock they cling to and allow them to flow freely into the well. Employing enhanced oil recovery methods requires capital investment and risk appetite, as these methods are not always successful.

Over the decades since the world started consuming oil in earnest, finding deposits has become more difficult: more enhanced recovery techniques are needed, deeper wells in more challenging locations must be drilled. Today, some 900 offshore rigs and mobile drill ships dot the globe, drilling in ocean waters up to 2 miles deep, and then up to 4 miles into the earth.

Oil exploration is highly risky, and thus is typically done as a partnership between two or more oil exploration and production (E&P) companies. The partnership first bids on a mineral rights lease from the host country for the right to invest in oil exploration in a given area. (Governments typically own the rights to underground natural resources, regardless of what private entity might own the land.) The national government chooses whether to allow drilling, determines the price it wants to charge, and signs either a royalty agreement or a production-sharing contract with the E&P firm.

Saudi Arabia has by far the most extensive oil reserves in the world. Canada recently vaulted to #2 when many industry players decided to officially recognize as "proven" Alberta's previously inaccessible Athabasca tar sands. The U.S., sitting atop just 2% of world oil reserves, is the world's most voracious oil consumer, followed closely by the European Union countries. In terms of production, the Organization of Petroleum Exporting Countries (OPEC) dominates, with 40% of world output. OPEC members are Iran, Saudi Arabia, Kuwait, Venezuela, Qatar, Libya, Indonesia, UAE, Algeria and Nigeria; Iraq is technically still a member, but it's post-Saddam governments have thus far generally disregarded OPEC export quotas. Non-OPEC countries like Russia, the U.S. and China also contribute a large portion of production.

As a large producer but even larger consumer of oil, the U.S. is the world's largest importer of oil. In 1994, the U.S. first became a net importer – the inevitable result of decades of declining production from mature oilfields. Despite the conventional wisdom that Americans are overly dependent on Middle Eastern oil, the U.S. imports only about 30% of its oil from that region. Forty percent of our oil comes from nearby Venezuela, Mexico and Canada; another 20% is imported from Africa, the North Sea region, and Russia.

Largest proven oil reserves		Top oil producers		Top oil consumers	
Saudi Arabia	22%	Saudi Arabia	13%	U.S.	25%
Canada	15%	Russia	12%	EU	20%
Iraq	9%	U.S.	8%	Japan	8%
UAE, Kuwait, Iran	8% ea	China	5%	China	8%

The U.S. imports about 30% of its oil from the Middle East.

Volts, Watts, and Joules

Not many people can keep straight the difference between volts, amps, watts, ohms, and joules. We all learned these concepts in high school physics, but even ex-engineers working in the electric power business have still been known to get confused. Nonetheless, having a rudimentary grasp of how electricity is measured can come in handy during interview discussions.

The classic analogy used to teach electrical quantity measurements is as follows: Imagine that you are standing in your backyard, holding a garden hose that you have pointed at a water wheel (don't bother wondering why you happen to have a water wheel in your backyard). The diameter of your hose is the resistance in this scenario, measured in terms of ohms. The force with which water sprays out of your hose is analogous to electrical voltage. Similarly, we refer to the speed with which water sprays out as amperage. Now, how big a water wheel can you turn with this spray of water from your hose? The size of the water wheel reflects the power or wattage of your hose setup. So, you stand there and spray water at the wheel so it spins 10 times – that is the amount work you are doing, measured in terms of joules.

Visit Vault at **www.vault.com** for insider company profiles, expert advice, career message boards, expert resume reviews, the Vault Job Board and more.

VΛULT CAREER LIBRARY   25

	Ohms	Volts	Amps	Watts	Joules Watt-hours
Quantity measured	Resistance	Pressure	Current i.e. flow of electrical charge (volts / ohms)	Power i.e. the flow of energy (volts x amps)	Work i.e. the flow of energy over time
Imagine a garden hose pointed at a water wheel...	Diameter of the hose	Force of the water coming out of the hose	How fast the water flows out of the hose	How big a water wheel the water stream can turn	How many times it actually turns the water wheel

To what extent are these measures applicable to everyday work in the energy business? First, people talk about the voltage of power lines. For the sake of minimizing transmission losses, when electricity comes out of the generator (25 kilovolts), it is "stepped up" by the transmission system operator to a high voltage (e.g., 230 kv, 345kv, 765kv), and then reduced again before entering a home or business (reduced to about 110 volts in North America, or about 220 volts elsewhere). Similarly, natural gas flowing in pipelines is compressed to a high pressure in large, interstate pipelines, but reduced to a safer pressure in the local distribution pipes that run underneath your house.

Secondly, people talk about the wattage of power plants. Power plants with a greater wattage literally have larger turbines (the water wheel in our backyard hose analogy). Power plants can operate at full or partial load – in other words, at 100% or less of their capacity. Multiply how many hours a plant operates by its wattage level, and you get the actual work output of the plant in terms of watt-hours. (Scientists would use the term "joules.") Note that wattage is the power input of a lightbulb, not the light output (which is measured in lumens): a 25-watt fluorescent bulb produces much more light than a 25-watt incandescent bulb, by using the same amount of power more efficiently (i.e. more of the work done is in the form of light rather than heat) to produce a brighter light. Your household electric bill indicates how many kilowatt-hours or electricity your lightbulbs and appliances used over the month. With respect to cars, semantic convention dictates that we refer to power or capacity as horsepower, which is completely synonymous with wattage.

The most crucial concept to make sure you keep straight is watts versus watt-hours. A power plant's capacity is measured in megawatts (millions of watts), whereas its actual output is measured in megawatt-hours. We talk about power plants producing a given number of megawatt-hours of energy in a year. Don't make the mistake of referring to something nonsensical like "megawatts per year"!

> Power plant capacity is measured in megawatts; power plant output is measured in megawatt-hours.

The Electricity Market

The electricity market is a fascinating example of classic microeconomics. If it isn't used as a textbook example very often, it's only because outsiders and generalists have a difficult time understanding the details of this complex market. Just like the price of any other commodity, the price of electricity in a given market (also known as the "market-clearing price") is set by the most expensive, or marginal, unit of supply.

What does the supply curve look like? An electricity supply curve is also known as a "stack": the rank order line-up of all power plants in a market from least to most expensive variable cost. (See Figure 1.4) Power plant operators bid their plants into the market in a reverse auction. (Multiple sellers bid to sell electricity and the winner of the auction is the seller who offers the lowest price. The winning price becomes the market price, and all power plants that bid less than the winning price are "dispatched" or told to turn on and generate power.) Economic theory suggests that most of them should bid in at their variable cost: if they bid less than that and get dispatched, they lose money by operating; if they bid more than that and don't get dispatched, they lose the opportunity to earn operating profit.

The dispatch curve or supply curve is fairly constant for any given market, only shifting as (1) plants are taken out of service for planned or unexpected maintenance, (2) as fuel prices change and consequently change relative operating costs, or (3) as new plants are built and enter the mix. The efficiency and fuel source of a power plant determines where it sits in the stack: the deeper a plant is in the stack, the more often it will be dispatched. The marginal or "swing" plant exactly covers its variable costs, while those deeper in the stack realize a positive operating profit that contributes to covering their fixed annual costs. (See Figure 1.5)

In most U.S. regional electric markets, gas-fired steam plants set the market-clearing price in most hours of the year. (However, when natural gas prices are very low, coal plants can become the marginal capacity.) Moving down

Visit Vault at **www.vault.com** for insider company profiles, expert advice, career message boards, expert resume reviews, the Vault Job Board and more.

VAULT CAREER LIBRARY   **27**

the stack from the gas steam plants, one then finds gas-fired combined cycles, coal plants, nuclear plants, and finally the hydro, solar and wind power plants, which have essentially zero variable costs. Solar and wind plants are known as "must-run" facilities, because they cannot actually be dispatched – none of us has control over when the wind blows or when the sun shines, so when it does, the output of those plants is incorporated into the market. Above the gas steam plants in the stack, one finds peaking plants that operate during peak demand hours with a variety of cost and response time profiles.

What does the demand curve look like? The demand curve shifts from left to right as aggregate market demand for electricity rises and falls over the course of a day, week, or season. Most electricity markets have a summer peak, brought about primarily by heavy air conditioning use; markets with high usage of electric heating also have a winter peak. The "peakiness" of a market (the ratio of peak annual demand to average demand) also varies from one region to another, reflecting the idiosyncrasies of electricity usage habits. A load duration curve summarizes the demand profile in a given market, depicting the frequency of demand at various levels. (See Figure 1.6)

Demand for electricity is fairly inelastic, meaning the demand curve is essentially a near-vertical line. While you are not willing to turn off your air conditioner and electric stove on a hot August afternoon when the utility's cost of power goes through the roof, some large industrial users of electricity are. Most notably, aluminum smelters that use electric furnaces often have deals with their power suppliers to shut down (and thus greatly reduce aggregate market demand) when market prices are high – they give up control over their own operations in exchange for a lower rate on the electricity they do use.

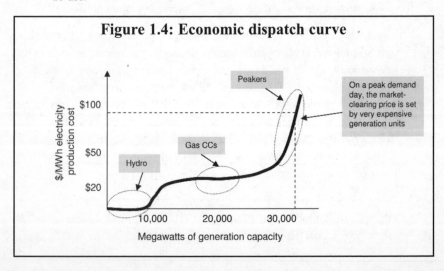

Figure 1.4: Economic dispatch curve

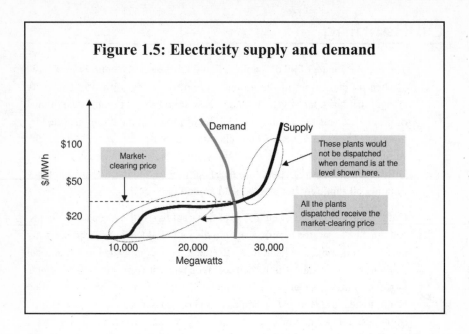

Figure 1.5: Electricity supply and demand

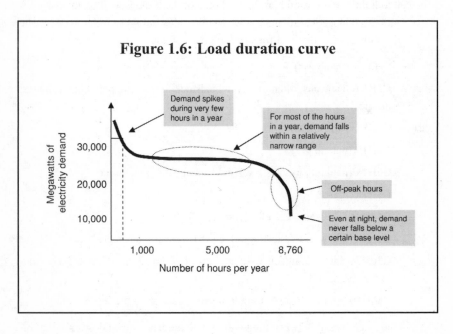

Figure 1.6: Load duration curve

Visit Vault at **www.vault.com** for insider company profiles, expert advice,
career message boards, expert resume reviews, the Vault Job Board and more.

VAULT CAREER LIBRARY

29

Oil Refining

The popular image of oil as a golden liquid turns out to be entirely misplaced – when oil comes out of the ground, it is often in the form of a chocolaty sludge, stiff black tar or greasy sand. These solid forms of oil must be diluted or processed near the wellhead into a liquid form that can be transported by pipeline. The resulting crude oil is classified in terms of its geographic origin, sulfur content ("sweet" = low sulfur, "sour" = high), and viscosity ("light", "medium" or "heavy"). Based on these definitions, there are some 200 types of crude oil routinely bought and sold around the world.

Crude oil is composed of hundreds of different types of hydrocarbon chains, and must be separated at refineries into distinct products in order to have any useful commercial applications. Refining is accomplished by fractional distillation: as crude oil is heated, components with varying boiling points vaporize and are separated out sequentially from lightest (shortest carbon chain) to heaviest (longest carbon chain). The heavier products are then often re-processed in a catalytic cracker to forcibly "crack" their long carbon chains into lighter, more useful products. Some of the heaviest end products are re-processed in a coker to make coke, a solid industrial fuel similar to coal. The most commonly distilled fractions of crude oil are:

CH_3 (methane gas)

C_2H_6 (ethane gas, often used as a feedstock for making petrochemicals and plastics)

C_3H_8 (propane gas)

C_4H_{10} (butane gas)

C_6H_{14} (naphtha, used as a solvent such as in dry cleaning)

C_8H_{18} (gasoline)

$C_{12}H_{26}$ (kerosene, used for jet fuel)

$C_{16}H_{34}$ (diesel fuel, heating oil, electric power plant fuel)

$C_{36}H_{74}$ (lubricating oil that is solid at room temp, often cracked into lighter products)

$C_{80}H_{162}$ (tar and waxes used to make asphalt, also made into coke)

The price spread between crude oil, wholesale gasoline, and heating/fuel oil is known as the "crack spread" – the difference between the cost of crude oil coming into a refinery and the revenue from its main outputs. The crack spread fluctuates wildly with world commodity prices, but if it's high enough, a refiner can cover the cost of the refining process and thus have an incentive to operate. The crack spread is typically calculated as follows:

		$ Gallon ("gal")	x	42	gallons barrel ("bbl")	X 1 bbl
	Heating oil	$ Gallon ("gal")	x	42	gallons barrel ("bbl")	X 1 bbl
+	Gasoline	$ / gal	x	42	gal / bbl	X 2 bbl
−	Crude oil	$ / bbl				X 3 bbl
=	Crack spread	($)				

Renewable Energy

Oil and natural gas form over millions of years as decaying organic matter is slowly crushed underneath layers of sediment and rock. While people argue over exactly how much oil and natural gas remains beneath the ground, the amount of fossil fuel in the earth is by definition finite. In contrast to fossil fuels, renewable fuels are naturally replenished after we use them, and thus provide a virtually endless supply of energy:

- **Wind turbines** sited in locations with strong, constant wind produce electricity by the slow rotation of three long blades. Turbines are typically installed on hilltops or open plains, with each tower rising 200 feet into the air. Wind turbine output ranges from 500kW to 3 MW each, and installations range from a single turbine up to several dozen in one array. In the U.S., windpower has occasionally encountered opposition from people who dislike the visual impact of turbine towers; however, in Europe, people tend to find wind turbines aesthetically pleasing.

- **Hydroelectric power** plants harness the energy in falling water to turn turbines. Large hydro plants are built as dams in major rivers, and can have thousands of megawatts of generation capacity. But flooding for large

Visit Vault at **www.vault.com** for insider company profiles, expert advice, career message boards, expert resume reviews, the Vault Job Board and more.

VAULT CAREER LIBRARY   **31**

hydro projects may displace native peoples, and the dams disrupt river ecosystems. Small or "run of the river" plants (less than 10 MW) simply use flowing river water to turn turbines, and don't require a dam.

- **Solar thermal** power plants use large arrays of parabolic mirrors to concentrate sunlight more than 80-fold to heat water into steam that can turn turbines.

- **Photovoltaic (PV)** systems are often casually referred to as "solar," but operate very differently from solar thermal plants, using heat from sunlight to excite electrons and create a current. The energy-intensive manufacturing process for PV products offsets some of the energy savings from PV (and drives the capital cost up quite high).

- **Geothermal** power plants pump water deep into the earth, where volcanic heat turns it to steam that can run turbines back on the surface. Injection wells and effluent ponds scar the landscape to some degree; however, in Iceland, which relies 100% on geothermal power, effluent ponds have been turned into high-end health spas.

- **Tidal** barrages generate electricity in much the same way as hydroelectric dams, and can be as large as 250MW. As the tide in an estuary ebbs, flowing water turns underwater turbines. But similar to hydro plants, these underwater dams and turbines disrupt the aquatic ecosystems.

- **Biomass** plants operate by burning wood chips, sugar cane bagasse, corn cobs, or other organic waste to heat water and run steam turbines, and are often built adjacent to farming areas or paper plants where combustible organic byproducts are produced. Plant material can also be made into ethanol, an oil that can be burned just like petroleum. Biomass is composed primarily of carbon, and thus produces a lot of carbon dioxide and particulates when combusted. However, since the fuel material would have been burned or decomposed anyway without generating electricity, these plants reduce net pollution in the system.

- **Landfill** gas is methane gas produced by decaying organic material. This natural gas can be captured as fuel for a power plant, rather than left to vent to the atmosphere. While methane is a fossil fuel and produces carbon dioxide during combustion, capturing and utilizing it offsets an equivalent volume of mined gas that would otherwise have been combusted.

People debate about which of the above technologies truly merit the label "renewable." Purists focus primarily on windpower, which is not only at present the most cost-effective renewable technology, but also the one that many agree has the least net impact on the environment.

Renewable resources supply just around 5% of world energy usage. Why not more? First of all, most energy consumption is for transportation and heating, while renewable technologies are focused on electricity generation. In other words, renewables supply a healthy percentage (about 20%) of world electricity needs (primarily from hydroelectric plants), but the world's electricity needs are only one part of total energy consumption. Secondly, many renewable electricity generation technologies are more expensive than fossil fuel generation. Even in cases where renewables are cost competitive with non-renewables, new technologies have a difficult time displacing entrenched standards: regulations can create an unfair disadvantage, and the public can be skeptical and disapproving.

Hydrogen is in some ways the ultimate renewable fuel, if only we could figure out how to produce it without using up as much energy as the hydrogen in turn produces for us. For this reason, hydrogen doesn't typically appear on a list of renewable energy sources: it's an output of energy use. Once hydrogen is produced, it can be used as a power plant fuel, or used to power fuel cells or combustion engines in cars.

While hydrogen is the most abundant element in the universe (90% of all atoms are hydrogen), it doesn't exist in its elemental state on our planet. To access hydrogen as a fuel, we must first split it out of water or organic matter through one of two ways.

1. If we use fossil fuels to generate electricity to electrolyze water, or steam to reform natural gas into hydrogen, we simply displace pollution from the point where hydrogen is used to the point where it is produced. Using non-renewable fossil fuel to produce hydrogen makes the resulting hydrogen a non-renewable resource as well.

2. On the other hand, if we use renewable resources to generate electricity to produce hydrogen and then use the hydrogen as fuel, then the entire system is emission-free and renewable.

The United States government's energy policy has recently focused prominently on the hydrogen-powered car. However, the much-criticized American "drive here, pollute elsewhere" policy envisions using domestic coal reserves to produce electricity to make hydrogen, liberating the U.S., in part, from imported oil. If we create hydrogen using coal, we don't reduce pollution – we only shift it from the roads to the power plants. In contrast, Europe has chosen to dramatically increase renewable resources in its

Visit Vault at www.vault.com for insider company profiles, expert advice, career message boards, expert resume reviews, the Vault Job Board and more.

VAULT CAREER LIBRARY   33

electricity supply over the coming years. Renewably-generated electricity will produce hydrogen, which can be both a clean car fuel as well as an energy storage mechanism, enabling the electric grid to function with a high percentage of wind and solar generation.

> Hydrogen is the most plentiful element in the universe other than perhaps stupidity." (Hunter Lovins, co-founder, Rocky Mountain Institute)

Hedging

Energy producers benefit from hedging their output to protect themselves in the scenario where market prices for their output fall. Likewise, energy product consumers have an interest in hedging against price increases. In addition to price risk, energy companies also hedge operational risk (e.g. a plant goes down unexpectedly for repairs) and credit risk (e.g. counterparty non-payment).

An energy market participant can hedge using exchange-traded or over-the-counter (OTC) derivatives, long-term contracts, supply diversification strategies, or specialty insurance. Exchange-traded derivatives include futures, options on futures, and swaps of futures contracts:

Exchange-Traded U.S. Energy Commodity Derivatives

- Crude oil futures, options, and swaps
- Heating oil futures and options
- Gasoline futures and options
- Propane futures
- Crack spreads futures and options
- Coal futures and swaps
- Electricity futures, options, and swaps
- Weather futures
- Spark spread futures and options
- Natural gas futures, options, swaps, and basis swaps

Futures and options can be combined in innumerable ways to create a suitable hedge for a particular company's situation and risk tolerance. For example, if you had a need to purchase natural gas next month, but were concerned that the price might be high at that point, you could buy gas options now in addition to buying the actual gas in one month – the net cost of the gas to you would be capped at the strike price (i.e. the pre-set exercise price) of your options. Alternatively, you could buy futures, which would fix the cost of gas to you absolutely, saving you money if prices go up, but resulting in an opportunity cost of overpaying if prices decline. A swap, in turn, locks in an effectively fixed price for a future purchase by "swapping" the floating price risk for a fixed price from the counterparty. We have illustrated several of the most common energy commodity hedging strategies in Figure 1.7.

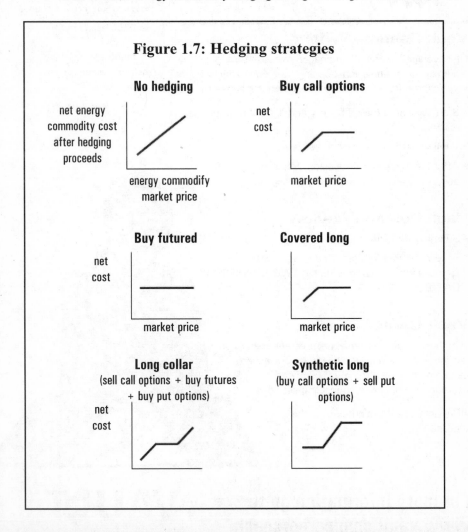

Figure 1.7: Hedging strategies

Visit Vault at **www.vault.com** for insider company profiles, expert advice, career message boards, expert resume reviews, the Vault Job Board and more.

VAULT CAREER LIBRARY   **35**

Major Industry Issues

In order to make an impression on higher-ups in the energy world, it's imperative that you grasp the issues that "keep them awake at night": Where are oil prices going over the long term? Are we running out of oil? How might pollution regulations change? Should I consider a potential carbon tax in my decision-making? Will changing regulatory rules of the game affect my business?

The answer to each of these questions can have a profound impact on a company's business strategy and profitability. Below are four issues that will give you a sense of the key industry topics you'll need to familiarize yourself with for interviews.

Oil Supplies

There is, by definition, a finite amount of oil in the earth. We will eventually run out of it. What is at question is when it will happen, and whether we will have enough advance warning to adapt. On one side of the debate, a variety of constituencies are concerned about the social and economic consequences of oil wells running dry, and are also interested in the environmental benefits of a transition to renewable sources of energy happening sooner rather than later. A proliferation of books with titles like The End of Cheap Oil, Out of Gas, and The Party's Over make that case passionately. On the other hand, incumbent oil interests are naturally optimistic about their own ability to find and extract oil, and tend to resist the argument that alternatives to their main product are in urgent need. Both sides of the debate do tend to agree that oil prices will shoot up when global production starts to permanently decline; estimates for when that will happen range from this year to 2100.

So, how much oil do we have left? One of the reasons this issue is so contentious is that oil supply is very difficult to measure. Oil supplies are categorized as follows:

- Undiscovered oil
- Possible reserves (discovered, with less than a 50% chance of being recovered)
- Probable reserves (discovered with greater than a 50% chance of being recovered)
- Proven reserves (recoverable)

Proven reserves are often overstated, as producers have every incentive to self-report a greater amount of wealth to their shareholders and lenders. Shell restated its proven reserves downward by 20% in early 2004, for example. OPEC countries have been in a "quota war" since the early 1990s, ratcheting their proven reserve statements up by impossible jumps of 50% at a time, in order to be allotted larger production quotas. Estimates of supply currently hover around 1 trillion barrels of proven reserves and some 2 trillion more ultimately recoverable out of probable, possible, and yet-to-be-found deposits. Contrast that with 30 billion barrels in exponentially accelerating annual worldwide consumption and, taking supply uncertainty into account, the problem quickly becomes apparent.

> Even oil companies agree that we will eventually run out of oil.

And what does an oil-less future look like? That depends how far distant that future is. In the near term, transportation can run off of natural gas (which is also finite, but substantially less depleted than oil), or electricity. Electricity can be generated by natural gas and coal, which by most accounts still exists in large enough quantities to fuel our society for a couple hundred years. In the longer term, we can envision the environmentalists' dream of a "hydrogen economy," in which renewable energy splits water molecules apart, and the resulting hydrogen powers efficient, zero-emission fuel cells.

Most oil companies have been investing in oil-substitute technologies since the mid-1990s (consider BP's "Beyond Petroleum" slogan, for example). Most of them also publish long-term forecasts of world energy consumption that show oil declining in importance over time relative to other current energy sources – those forecasts, however, tend to show a disturbingly substantial segment of future energy demand met by an unknown source labeled "surprise" or "future discoveries."

Conventional economic wisdom suggests that when oil supplies begin to wane, prices will rise, and demand will fall in response, thus extending the life of the remaining supplies. Rising prices would make technological innovation profitable, improving our ability to extract oil from small and remote deposits previously thought unrecoverable. Adherents to such a theory point to the oil shocks of the late 1970s, which motivated radically reduced consumption and investments in renewable energy technologies.

However, those who are concerned about an imminent oil shortage point out that the theoretical macroeconomic refrain "demand will fall" belies a lot of microeconomic pain and disruption. When oil prices increase, the cost of most everything one can buy increases, which forces us to consume less, accept fewer consumption choices, and make difficult lifestyle sacrifices. Rising oil prices slow the economy, drive down employment, and lower standards of living. So, just because the market will, at a high level, "adjust," doesn't mean that dwindling oil supply is not a crisis-level problem. Discovery of new oil fields peaked back in the 1960s, and oil production peaked for many countries (including the U.S.) as long ago as the 1970s. The fundamental question is: will we know with enough certainty far enough ahead of time to make the market "adjustment" a gentle one? And if not, doesn't it behoove us to start that transition now?

> The Stone Age didn't end because they ran out of stones; the Oil Age won't end because we run out of oil." (Don Huberts, CEO, Shell Hydrogen)

Electric Power Regulation

How an industry is regulated makes a profound difference in how participant companies operate, the decisions they make, the products they produce and prices they charge. In the case of the electric power sector, regulatory restructuring over the past decade has moved the industry away from a system of local utility monopolies toward a model of greater competition among generators and distributors of electricity. This restructuring movement is often imprecisely referred to as "deregulation"; however, it has been achieved by adding federal and state government regulations to the books, not by reducing the scope of government. Just like the "deregulated" airline and telecom industries, the electric power industry is still carefully regulated by a variety of agencies on the lookout for consumer and investor fraud, monopoly behavior, collusion, excess pollution, price gouging, worker safety violations and the like.

Prior to recent industry restructuring, the electric power sector consisted predominantly of utility companies holding monopolies to generate, transmit, and distribute electricity within a specific geographic area. Each utility took care of building enough power plants to supply its own customers, including enough excess capacity to ensure reliability. Utilities would also buy and sell power amongst themselves on a limited basis to balance the supply and demand in their own systems. State Public Utility Commissions (PUCs) approved the rate that each utility charged to its customers (households,

Visit Vault at **www.vault.com** for insider company profiles, expert advice, career message boards, expert resume reviews, the Vault Job Board and more.

VAULT CAREER LIBRARY   **39**

stores, factories) based on the amount the utility spent to build and operate its own plants and power lines.

The impetus to change this system came from arguments about market efficiency. With each individual utility responsible for carrying enough excess capacity to ensure reliability, the total amount of reserves across regions as a whole was more than necessary – and each utility's customers were paying for all of that excess capacity to sit idle, like expensive blackout insurance. Because utilities could recover the cost of their investments in pre-set, regulated retail prices, they had little incentive to tightly manage capital and operating costs or to put a lot of effort into technological innovation. Looking around at the rest of the developed world, this was one market where the United States was a laggard: the UK, Australia, Argentina, Chile and New Zealand all moved away from a monopoly system long before the U.S. did. Ironically, despite having a structure that fostered inefficiency, the U.S. electric power industry provided power at prices lower than in most of the rest of the world. However, the 1990s was a time when the notion of competitive markets was very much in vogue, so the "deregulation" argument was an easy one.

Regulatory restructuring of the electric power sector involved four different initiatives, which were implemented at the federal level, and to varying degrees at the state level. First, Congress passed the 1992 Energy Policy Act, which forced utilities across the country to buy power at fair rates from anyone who wanted to generate electricity in their service territories. This caused a boom in construction of "independent" power plants, built and operated by newly-formed private companies, or independent affiliates of utilities. Utilities for the most part stopped building much new capacity themselves, finding that it was cheaper to let these aggressive new players make the investment decisions. The role of electric power traders also came into being, with new utility subsidiaries and independent companies like Enron jumping into the business of buying and selling power across the country.

Then, during the mid- to late-1990s, individual state legislatures and PUCs addressed the goal of lower retail prices by enacting some or all of three regulatory changes:

- **Centralized wholesale market and plant dispatch.** Many areas of the U.S. created a nonprofit Independent System Operator (ISO) with operational control over all the generators in a region. New England states had recognized for years the cost savings from such resource-sharing, and the New England ISO was used as a model for others in the mid-Atlantic, California, New York, Texas and the Midwest. Most ISO regions also created a centralized wholesale market, into which all generators (including utilities) have to sell their power. Thus, instead of using bilateral contracts to buy power, utilities and any other load-serving entities buy power out of "the market." A central power exchange (PX) is a great equalizer, preventing large entities from having greater bargaining power to affect pricing, and ensuring small generators get fair treatment.

- **Asset divestiture.** A handful of states passed legislation forcing utilities to divest their generation assets (i.e. their power plants), breaking up their vertical monopolies. Theory suggests that independent owners of the assets would operate them at a lower cost, resulting in lower prices; splitting the assets into multiple hands would also increase competition and further drive down prices. When assets were sold at auction, the highest bidder was often another utility or its independent affiliate.

- **Retail access.** Many states legislated an end to utilities' geographic service territories by granting a universal right to sell retail electricity service. In those states, households and businesses can now choose their electricity provider just like they choose their long distance provider. In theory, the pressure to offer customers a low electricity price to retain their business creates the incentive for utilities to generate or procure power at the lowest possible cost. In many cases, however, retail prices were low enough to begin with, such that undercutting them has been difficult. In states with retail competition, very few households, and only some electricity-intensive businesses, have exercised their right to choose an alternate power provider.

> **Electricity industry restructuring was intended to provide lower prices to consumers.**

States/regions with the highest electricity prices started down the restructuring path first: notably New England, Pennsylvania, New York and California. Pennsylvania created an ISO and mandated retail access, and has since been generally hailed as a restructuring success story. Other regions, such as the Midwest, suffered capacity shortages and severe wholesale power price spikes as they struggled to design and implement detailed "rules of the game" for newly restructured markets. In 1998, California implemented an ISO, PX, asset divestiture and retail access, and just 3 years later suffered the

Visit Vault at **www.vault.com** for insider company profiles, expert advice, career message boards, expert resume reviews, the Vault Job Board and more.

V/\ULT CAREER LIBRARY   **41**

worst power crisis in U.S. history, becoming a poster child for anti-restructuring arguments.

Between December 2000 and March 2001, California wholesale gas and electricity prices spiked at up to 10 times their normal levels, and capacity shortages resulted in rolling blackouts. In parts of California where the local utility was allowed to pass procurement costs through to customers, consumer electricity bills tripled. In contrast, utilities that were forced to buy power from the volatile wholesale market, but sell it at pre-set regulated rates, suffered billions in losses. Pacific Gas & Electric, the country's then-largest utility, actually went bankrupt despite a large-scale bailout attempt by the state government. For several years after this crisis, the power sector was overwhelmingly focused on understanding its causes and trying to draw conclusions for the future of regulation.

The infamous California energy crisis of 2000/2001 is commonly attributed to three concurrent causes: (1) a "perfect storm" of market fundamentals, (2) poor regulatory design, and (3) illicit market manipulation:

1. **A perfect storm.** Hydroelectric plants in the Pacific Northwest normally provide one fifth of California's power; but in 2000, spring draughts substantially reduced hydro capability, compressing the supply curve. At the same time, natural gas bought cheaply in the summer and stored underground normally provides much of California's wintertime gas needs; however, because gas plants had to operate more in the summer to make up for lost hydropower, gas was expensive in the summer of 2000 and storage facilities could not be filled, further tightening supply. In addition, because power plants were running overtime to compensate for lost hydropower, they also went down for maintenance more often during the fall of 2000 and winter of 2001. Winter storms washed kelp into the intake system of one of California's main nuclear plants, forcing this backbone of the grid to operate at greatly reduced capacity at the same time as those winter storms caused electricity and gas demand to soar. On top of everything, capacity reserves in the state were low because nobody had built new power plants in years, due to uncertainty over whether and how restructuring would take place, as well as strong not-in-my-backyard ("NIMBY") opposition from residents. This unfortunate coincidence of events reduced supply and increased demand for gas and electricity, creating a textbook example of a constrained market with extremely high price spikes.

2. **Market structure problems.** Problems with California's newly restructured power market exacerbated the impact of high prices on the

industry and consumers. California mandated asset divestiture without vesting contracts for plant output, meaning that when supplies became constrained, the state had no authority to demand that power plants sell electricity to in-state buyers. While forced divestiture aimed to reduce generator market power, the state didn't go far enough with such safeguards. Some generators controlled up to 8% of the state's capacity, which turned out to be enough to exercise market power – i.e., in a constrained market, they could offer their output at extremely high prices and still make a sale. Observers noticed this type of market power gaming before the crisis peaked, but the Federal Energy Regulatory Commission (FERC, who has jurisdiction over monopoly behavior) refused to act. Most dramatically, the California legislature had mandated that utilities buy all of their power in a central spot market, while providing it to customers at fixed, regulated rates. When this rule was written, a scenario of wholesale prices exceeding regulated retail rates was considered inconceivable; however, as wholesale prices spiked in the winter of 2001, utilities found themselves buying high and selling low, massively subsidizing every kWh of electricity supplied to customers until they had no funds left to do so, and the government stepped in to procure power for the state's residents.

3. Manipulation. A constrained market and poorly designed regulations resulted in a situation ripe for both clever arbitrage and outright manipulation. In the first case, electricity generators and traders in California responded to the incentives that the legislature created for them with the new regulatory structure. For example, glaring loopholes in the restructured market rules allowed players to export and re-import power into the state to avoid price caps, thus driving prices higher. Similarly, players could profit by arbitraging pricing on both sides of a transmission bottleneck, making the bottleneck worse rather than alleviating it. However, in electricity markets there is a rather fine line between clever arbitrage and illegal actions – exercising market power is illegal, but whether a company possesses market power or not changes hour to hour as market supply and demand fluctuate. Clear illegal actions during the heat of the energy crisis included withholding generation capacity to drive up prices: declaring false outages, failing to bid output into the PX, bidding output into the PX at unjustifiably high prices. Generators also illegally gamed the system by submitting false load schedules, double-selling the same electricity, selling nonexistent electricity, deviating from dispatch instructions, and colluding with other generation companies to play these profiteering games.

Visit Vault at **www.vault.com** for insider company profiles, expert advice, career message boards, expert resume reviews, the Vault Job Board and more.

VAULT CAREER LIBRARY   **43**

Years of public investigation into the causes of the energy crisis focused overwhelmingly on market manipulation games. Ultimately, the state of California and its residents received several billion dollars in fines from the biggest culprits: the Williams Companies, El Paso Corporation, Dynegy, Reliant Resources, Enron, Duke Energy, Mirant Corporation. Scrutiny of Enron's immense profits in California during the winter of 2001 helped draw attention to its web of partnership hoaxes, resulting in the company's astonishing crash into bankruptcy in the fall of 2001. In 2002, Congress passed the Sarbanes-Oxley Act in response to the Enron debacle, mandating accounting oversight boards for public companies, requiring companies to furnish enhanced financial disclosures with personal accountability, and providing stiffer sentencing guidelines for white-collar crime.

The rest of the country watched the California energy crisis in horror, with most observers finding a lesson about why electricity market restructuring should be halted or re-examined. In the fall of 2002, FERC issued a proposal for a carefully-formulated standard market design that could be implemented across the entire country at once, rather than let each individual state wrestle as California did with internal political lobbies to create a flawed regulatory regime. However, FERC's proposal was ripped apart from all sides by constituencies that had different ideas of what a fair market would look like. Since 2001, almost all states have simply frozen their restructuring activities in place. As a result, states now sit at very different points along the continuum between local regulated monopolies and competitive markets for generation and distribution. With those markets for the most part functioning well despite their heterogeneity, the electric power industry expects to simply remain regulated as it is for the foreseeable future.

> The California energy crisis was caused by market fundamentals, flawed regulatory design, and illicit market manipulation.

Climate Change

When sunlight hits the earth's surface, some of it bounces back into the atmosphere, where it is absorbed by atmospheric carbon dioxide (CO_2) and water vapor, preventing heat from escaping the atmosphere into space. Were it not for this "greenhouse effect," the earth would be uninhabitably cold. However, human activity since the industrial revolution has artificially added enormous amounts of carbon dioxide and other "greenhouse gasses" (GHG) into our atmosphere. These anthropogenic emissions appear to be increasing the greenhouse effect enough to significantly affect global climate patterns. Unfortunately, the energy sector is the primary culprit: electricity generation is responsible for one third of the increased GHG concentrations, and fuel use for transportation is responsible for nearly another third (the remaining third results from a combination of deforestation and manufacturing processes).

During the 20th century, the earth warmed by an average of 0.6° Celsius, with portions of the Arctic and Antarctic warming by 5° C. Melting of polar ice caps resulted in an average sea level rise over the century of 6 inches. At current rates of increase of GHG-generating activity, climate models predict an additional rise in average temperature of up to 14° C by 2100, and a host of changes to weather patterns:

- **Increased average temperature.** Air and water temperature changes would be distributed unevenly around the globe. Moderate warming in the U.S. would be a boon to agriculture, but would also expand the tropics and promote increased spreading of infectious diseases. Given enough time, animal species can adapt to climate changes, but if climate change occurs quickly (like, within a couple hundred years), many species would likely die off.

- **Increased average precipitation.** Precipitation changes would vary regionally, and be offset to an uncertain degree by increased evaporation. Dry areas are predicted to get drier, putting drinking water supplies and agricultural viability in question; coastal areas would be expected to receive more rainfall, increasing flood risk.

- **More chaotic weather.** Higher ocean temperatures would lead to increased frequency and severity of storms.

- **Localized cooling effects.** For example, as the Arctic ice cap melts, cold freshwater is entering the North Atlantic. An increase in this phenomenon could be expected to stagnate the Gulf Stream and dramatically cool northern Europe.

- **Higher sea level.** Climate models predict a sea level rise of up to 35 inches by 2100 if current GHG emissions rates remain unchecked and polar ice cap melting accelerates. Coastal flooding in low-lying areas like Bangladesh, the South Pacific islands, and the Netherlands would increase dramatically in such a scenario. Even if temperatures stabilize at 3° C above current levels, ice cap melting would continue slowly for the next thousand years, ultimately increasing sea level by 20 feet.

Visit Vault at **www.vault.com** for insider company profiles, expert advice, career message boards, expert resume reviews, the Vault Job Board and more.

VAULT CAREER LIBRARY   45

What elements of such climate change predictions are in question? Principals of chemistry tell us that greenhouse gasses have a warming effect. Historical records detail the enormous amounts of greenhouse gasses that human activities have added to the atmosphere. We can directly observe that the earth has indeed warmed during the last century. What cannot be known for certain is whether the correlation of these phenomena reflects causality. In the 1970s, atmospheric scientists proposed a theory of climate change to describe a causal link. This theory has steadily gained adherents, and today there is broad consensus in the international scientific community that climate change is resulting from human GHG emissions.

Like all scientific theories, climate change can never be proven. Accumulating evidence can only make a theory more widely accepted. Nonetheless, we structure our lives based on the acceptance of many scientific theories: germ theory provides the rationale for developing medicines, atomic theory allows us to make nuclear bombs, gravitational theory makes us wary of walking off cliffs. Climate theorists advocate adoption of the precautionary principle in response to the threat of human-induced climate change: act now to prevent climate change because by the time the amount of evidence approaches irrefutability, the devastating effects would have already occurred.

The climate change question is whether, when, and by how much we will choose to reduce our GHG emissions and thereby prevent some amount of the devastation forecasted by today's climate models. Happily, reducing GHG emissions often goes hand-in-hand with increasing energy efficiency and thus saving money, which is why countries and companies around the world began voluntarily reducing emissions many years ago, despite any lingering uncertainties about climate change theory.

Carbon dioxide is the most prevalent greenhouse gas. However, methane and nitrous oxide in the atmosphere also trap heat. In addition, a number of purely man-made substances have extreme greenhouse properties:

Greenhouse gas	Atmospheric concentration % increase since 1800	Primary anthropogenic sources	Warming potential relative to CO2
Carbon dioxide (CO2)	37%	Fossil fuel combustion	n/a
Methane (CH4)	200%	Landfills, livestock	21x
Nitrous oxide (N2O)	15%	Fertilizer use	310x
Fluorine-containing gases (SF6, CF4, HFC, PFC)	n/a (man-made only)	Aluminum smelting, electrical insulation	Up to 24,000x

The United States is responsible for 25% of global CO2 emissions. As a result, the U.S. has been at the heart of the climate change debate, annually producing 4.5 tons of carbon emissions per capita, compared to 2.3 tons per capita in Europe and just 1 ton per capita worldwide.

> The energy sector is responsible for some 60% of the emissions that are thought to drive climate change.

In 1997, representatives from 160 countries collaboratively proposed a solution to the greenhouse gas emissions problem: the Kyoto Protocol to the United Nations Framework Convention on Climate Change. This legally binding treaty sets a year 2012 cap on GHG emissions for each industrialized country that chooses to ratify it. The Kyoto Protocol finally came into force in 2004, when Russia became the 126th country to ratify it, bringing the percent of global GHG emissions represented by the ratifiers up to the 55% requirement.

Notably absent from the list of Kyoto ratifiers has been the United States (in addition to coal-rich Australia, the only other major holdout). In the years right after the Protocol was written, the Republican-controlled Senate announced it would not ratify the treaty if then-President Clinton submitted it for consideration; and in more recent years, President Bush has been quite clear that he will not support it (despite a 2004 Pentagon study emphasizing the national security concerns associated with ignoring climate change). Notwithstanding the administration's feelings on the topic, bipartisan legislation to provide incentives for voluntary reductions, require disclosure of emissions levels, and impose a cap on companies' emissions continues to circulate through Congress. In addition, several individual U.S. states have enacted policies to specifically cap carbon emissions from power plants.

Despite the criticism leveled at the U.S. for being the largest GHG producer yet refusing to ratify Kyoto, it ultimately may not matter whether our government adopts the protocol or not. U.S. companies decided years ago that voluntarily capping their carbon emissions would be in their own best interest. Today, there is a Chicago Climate Exchange, where carbon credits are traded as a commodity, just like corn futures are traded at the Chicago Mercantile Exchange. But why would a company voluntarily reduce its carbon emissions? Reasons include:

- **Legality.** Divisions and subsidiaries of U.S.-based firms operating in Kyoto-participating countries are bound by their host country's laws to meet Kyoto targets. If a company goes through the effort to identify

Visit Vault at **www.vault.com** for insider company profiles, expert advice, career message boards, expert resume reviews, the Vault Job Board and more.

VAULT CAREER LIBRARY   **47**

emissions-reduction strategies in its international offices, it may very well find it prudent to go ahead and make similar changes to its domestic operations.

- **Long-term planning.** Many people believe that, despite the current administration's stance on the issue now, one way or another, the U.S. will eventually impose a carbon cap. Some companies find it fiscally prudent to take action now in order to be prepared, and possibly get retroactive credit for early action.

- **Influence.** Companies may believe that those who establish a strong track record in voluntary carbon reductions will be more credible participants in defining the particulars of whatever carbon legislation is one day adopted in the U.S.

- **Profit.** The global carbon credit market is becoming a substantial new trading venue for companies. With huge demand for credits abroad, reducing one's own emissions to create saleable credits has become an investment opportunity for U.S. firms. In addition, many emissions-reduction solutions are money-savers in their own right, and serve to reduce enterprise risk.

- **Innovation.** Companies committed to reducing their carbon emissions will rely on new technologies and innovative efficiency measures to do so. If U.S. firms simply "sat out" from this process, they could fall behind on certain technological advances, putting their competitiveness in the global marketplace at risk.

- **Employee retention.** Companies are increasingly sensitive to their employees' desire to work for the "good guys." In fact, employee initiatives initially drove some of the major oil companies to embrace the climate change issue years ago.

- **International pressure.** Generally, operating as a pariah in one's industry in the face of constant criticism is not a pleasant position in which to be. U.S. firms who choose to embrace environmental stewardship – whether it be in terms of climate change or other types of pollution – tend to have an easier time doing business abroad.

- **Ethics.** Corporate officers are bound by law to put their shareholders' interests first. However, they do have leeway in how to define or meet such a goal. Corporate directors are, after all, people who stand to be adversely impacted by a warming climate just like everyone else … and most of them are thus fundamentally interested in understanding climate change science and doing their part to work towards a solution.

While current carbon reduction activity in the U.S. is entirely voluntary at present, energy companies do believe that a carbon tax of some sort will eventually exist. Utilities and lenders routinely conduct sensitivity analysis on their assets by incorporating a cost assumption for carbon dioxide emissions.

> Reducing greenhouse gas emissions has proven to be costless for many major corporations.

How will action on the climate change issue affect the energy sector? We can expect to see an ever-stronger push to develop renewable energy generation, implement low-emission coal technologies, further develop relatively low-carbon natural gas resources, improve energy efficiency of the industrial sector, increase fuel economy of combustion engines, and potentially even re-invigorate the nuclear power industry. Climate change is a complex problem whose solution will ultimately come from a variety of sources – for bright, technology-savvy people interested in problem-solving, the energy sector thus offers a wealth of opportunity for meaningful work.

Pollution

Fossil fuels provide the vast majority of our energy – we burn them to drive cars, to heat our homes, run our factories, and to make electricity. Unfortunately, fossil fuel combustion results in substantial air pollution: smog, soot, acid rain, and greenhouse gases. We have known this ever since the Middle Ages, when the use of coal for home heating saw ceilings turn black, children cough and wheeze, and air become smoggy. However, only since the 1970s have industrialized countries made a concerted effort to reduce pollution from fossil fuel combustion.

Combustion is the rapid combination of fuel with oxygen (slow combination with oxygen is known as oxidation, like rust forming on your car). As a hydrocarbon's carbon and hydrogen atoms rearrange themselves into carbon dioxide and water, the formation of new chemical bonds releases energy, which we can harness as heat or force to do work:

Visit Vault at **www.vault.com** for insider company profiles, expert advice, career message boards, expert resume reviews, the Vault Job Board and more.

VAULT CAREER LIBRARY   49

Theoretical combustion equation for methane
$CH_4 + 2O_2 => CO_2 + 2H_2O + 890kJ$ energy

If we could combust natural gas with pure oxygen, the process would indeed produce just carbon dioxide, clean water, and energy. However, we burn natural gas in ambient air, which contains only about 20% oxygen, a lot of nitrogen, and a host of trace elements. The high temperature of combustion causes the nitrogen and other elements in the air to reformulate into unsavory substances. In addition, natural gas is not really pure methane, but usually contains other trace elements, such as sulfur. As a result, burning natural gas produces a number of air contaminants in addition to carbon dioxide.

The situation is worse with coal and oil, which are far more complex hydrocarbons than natural gas. Burning coal and oil produces substantially more carbon dioxide and nitrogen and sulfur compounds than natural gas (See Figure 1.8), plus a host of carcinogenic and neurotoxic heavy metal particulates.

Theoretical combustion equation for methane	
Natural gas:	CH_4
Coal:	$C_{135}H_{96}O_9NS$ (plus trace amounts of As, Pb, Hg)
Oil:	C_4H_{10}, C_6H_{14}, C_8H_{18}, $C_{12}H_{26}$, $C_{16}H_{34}$, $C_{36}H_{74}$, etc (plus trace amounts of S)

In 1970, Congress passed the landmark Clean Air Act (CAA) – a first step towards addressing rampant air pollution from power plants, cars and factories. The law was then updated in 1977, 1990 and 1997. The Clean Air Act essentially requires power plants to apply for a pollution permit. Each plant is then granted an amount of allowable emissions; if it chooses to exceed that amount, the company must purchase additional emission "allowances" (or "credits") on the market. Though the CAA is theoretically enforced by the Environmental Protection Agency, in reality it is not often enforced at the federal level, and individual states are left to fine and sue non-compliers.

One crucial gap in the Clean Air Act is the treatment of coal plants. For the most part, existing coal plants were exempted from CAA rules from the outset, because regulators believed that old coal plants would gradually be replaced by newer, cleaner plants anyway. However, those expectations were wrong – hundreds of dirty, 50-year-old coal plants continue to operate around the country. Pursuant to 1990 CAA amendments, these grandfathered plants would nonetheless become subject to CAA emissions standards if they were upgraded or expanded by their owners. However, this "new source review" provision was weakly enforced, and effectively eliminated by the White House in 2003. In March 2005, the EPA issued the Clean Air Interstate Rule (CAIR), which mandates steep reductions in state-level SO2 and NOx emissions in the eastern half of the US, to be phased in through 2015; with this rule, it is up to the individual states to determine how to meet the emissions limits, whether by choosing to clamp down on their old coal plants, or by other measures.

Let's take a look at each of the major sources of air pollution created by the energy sector:

Nitrogen oxides

Nitric oxide (NO) and nitrogen dioxide (NO2) are gases formed during combustion when high temperatures cause oxygen and nitrogen in ambient air to reformulate. About 55% of manmade NOx (pronounced "nox") emissions come from cars and vehicles, and another 30% come from power plants. (Don't confuse NOx with nitrous oxide (N2O, a.k.a. "laughing gas"), which is a greenhouse gas by-product of automobile catalytic converters.)

Nitrogen oxides are harmful in two primary ways: they react with sunlight to form ground-level ozone (O3, a.k.a. photochemical smog, or brown smog); and they react with water in the air to make nitric acid (HNO3), or acid rain. Man-made ozone triggers 6 million asthma attacks each year, along with long-term lung tissue damage leading to increased frequency of bronchitis and pneumonia. NOx emissions create acid rain hundreds of miles away from their source, acidifying some rivers in the Northeast to the point where all fish life is destroyed, as well as killing trees and corroding buildings.

Power plants can control NOx emissions with selective catalytic reduction (SCR) systems, which use ammonia to transform nitrogen oxides back into harmless nitrogen and oxygen. Cars use catalytic converters for the same purpose. Power plants can also use low NOx burners, which burn fuel in stages and at lower temperatures to prevent NOx formation in the first place.

In 1990, the EPA instituted a novel "cap and trade" system for nitrogen oxide emissions from power plants. Each plant can decide whether to invest in emissions reduction technology to meet emissions guidelines, or to purchase

Visit Vault at **www.vault.com** for insider company profiles, expert advice, career message boards, expert resume reviews, the Vault Job Board and more.

VAULT CAREER LIBRARY   **51**

emissions allowances in an open market. Plants that have access to capital or the technological capability to cheaply reduce emissions below their allowed level can actually make money by selling their unused allowances to other power plants – overall, the total amount of emissions in the system is capped, and the cost of meeting that cap is minimized. NOx credits trade for around $2500 per ton, and a typical new natural gas plant with SCR might spend a couple hundred thousand dollars per year on purchased credits. A new coal plant, in contrast, would likely pay a couple million dollars per year (due to its higher emissions rate) if all of its emissions were subject to regulation.

Sulfur dioxide

Sulfur dioxide (SO2) is a gas formed during combustion of sulfur-containing fuels. The vast majority of SO2 emissions come from coal- and oil-burning power plants and petroleum refineries. Sulfur dioxide is harmful in two primary ways: it reacts with the atmosphere to form soot, or gray smog; and it reacts with water in the air to make sulfuric acid (H2SO4), or acid rain. Gray smog causes lung disease, and has been a high-profile killer in urban areas ever since industrialization: for example, in 1952, one of London's infamous pea soup smogs killed 4,000 people in one week; in 1948, half the residents of one town in Pennsylvania died or were hospitalized when coal smog turned the midday sky black.

Power plants primarily mitigate SO2 emissions by adding a flue gas desulphurization mechanism (FGD, or "scrubber"), which traps sulfur compounds in the emissions stream before they enter the atmosphere. Newer coal plants sometimes use fluidized bed combustion (which employs limestone dust to absorb sulfur) or coal gasification (where coal is converted into methane by the addition of hydrogen, decreasing the ratio of carbon and sulfur to energy output).

As with NOX, SO2 is also regulated by a cap and trade program. SO2 allowances sell for about $700 per ton. Due to their extremely low sulfur emissions, new gas plants only spend a couple thousand dollars per year on credits; the newest clean coal plants, in contrast, pay a few hundred thousand. Old coal plants would owe tens of millions per year — if they were not exempted from SO2 emissions regulations.

Mercury

Mercury is one of the more recently understood pollutants. Though children at one time used to play with the liquid metal, mercury is in fact highly toxic to the nervous system, causing mental deficiencies and birth defects. It tends to bioaccumulate in fish, from which we then ingest it in concentrated doses. Estimates differ on what percent of environmental mercury emissions come from coal power plants (up to about 40%), versus from medical waste and car incineration or from consumer products.

Power plants can mitigate mercury emissions by injecting absorbent activated carbon into flue gases. However, the specific profile of the best mercury mitigation technology can differ from plant to plant, based on particular characteristics of the flue gas temperature, pressure, and composition. In March 2005, the EPA issued the first-ever mercury emissions regulation, which will ultimately reduce total mercury emissions from coal-fired power plants by 70% across the country. As of this writing, debate is still ongoing as to the details of an enforcement mechanism, and state governments are beginning their analyses of whether to simply enforce the EPA regulation as is or issue a more stringent one.

Particulates

Particulate emissions are a mixture of microscopic solids (metals, soil, dust, allergens) and liquid droplets (water, acids, organic chemicals) suspended in air. About one third of man-made particulates are produced by vehicles (primarily diesel ones), another third by power plants (primarily coal ones), and another third from routine household activities and road dust.

While particulates many not be the fanciest pollutant in terms of chemistry, they have gradually become recognized as one of the most serious. Fine particles of soot in the air cause asthma, heart arrhythmias, heart attacks, and lung cancer. Particulates are now known to be directly responsible for approximately 64,000 premature deaths per year, or double the number of deaths from car accidents.

Diesel cars and trucks use an oxidation catalyst or simple fabric filter to remove particulate matter from their emissions streams. Similarly, coal power plants use baghouses (large fabric filters) or electrostatic precipitators, which pull particles out of flue gases using an electric charge. In 1997, the EPA tightened its regulations on particulate matter to cover suspended solids as small as 2.5 microns in diameter (compared to the previous 10 micron limit).

Visit Vault at **www.vault.com** for insider company profiles, expert advice, career message boards, expert resume reviews, the Vault Job Board and more.

VAULT CAREER LIBRARY   **53**

Carbon monoxide

Carbon monoxide (CO) forms when inadequate oxygen is available in combustion to form carbon dioxide. This problem of incomplete combustion occurs primarily in cars and household appliances. Carbon monoxide is well-known as a suicide aid, but in lower doses in the atmosphere, it causes fatigue and contributes to heart problems over time.

Catalytic converters help prevent CO formation, in addition to their primary function of nitrogen oxide reformulation. In addition, gas stations in many states now sell oxyfuel – gasoline enriched with extra oxygen — to ensure sufficient supply for complete combustion.

Carbon dioxide

Carbon dioxide (CO_2) is a large volume product of hydrocarbon combustion. About 40% of human CO_2 emissions in the U.S. come from power plants, and another 25% from cars. Carbon dioxide is a greenhouse gas that contributes significantly to ongoing anthropogenic climate change. While companies are experimenting with carbon sequestration techniques to remove it from exhaust streams, the only current commercially viable mitigation options are to (1) increase combustion efficiency and thus use less fuel per unit of energy output, or (2) not use fossil fuels. The U.S. presently has no regulations in place for carbon dioxide emissions from any source. However, recent adoption of the Kyoto Protocol in most major industrial countries may increase voluntary compliance in the U.S. and put pressure on it to eventually implement emissions restrictions.

Proposals for the updating of air pollution regulations abound in Congress, the White House, industry lobbies, consumer groups, and environmental advocacy organizations. The most anti-environmental positions favor creating new standards for all plants that would be weaker than current CAA standards, across the board. More pro-environment groups have proposed simply enforcing the existing CAA, which would result in a significant incremental reduction in emissions.

The debate is contentious, with utilities lobbying in earnest to play for time. As a result, industry observers generally expect to see tighter national-level air emissions controls materializing in the 2012 timeframe or later. In the meantime, many individual states have taken matters into their own hands, and gone forward in implementing stricter emissions rules to protect their citizens' health. The New England states, in particular, have implemented very strict sulfur dioxide and nitrogen oxide regulations, and are in the process of working on carbon dioxide standards.

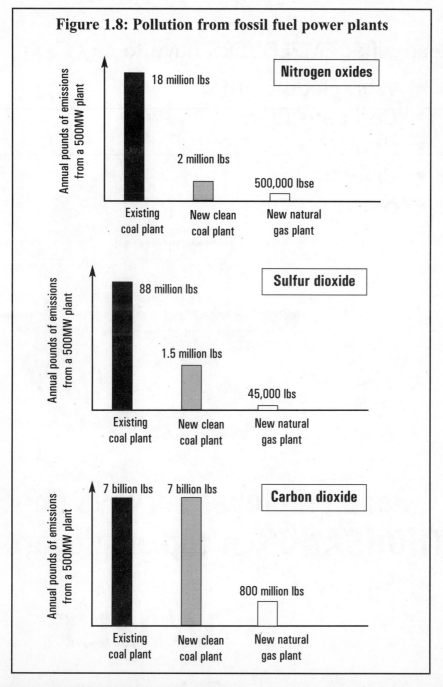

Figure 1.8: Pollution from fossil fuel power plants

Visit Vault at **www.vault.com** for insider company profiles, expert advice,
career message boards, expert resume reviews, the Vault Job Board and more.

VAULT CAREER LIBRARY

55

GETTING HIRED

Energy Industry Job Opportunities

CHAPTER 4

Which Job Function?

In order to pursue a job in the energy sector, your first decision is what type of position you want – in other words, what functional role you want to play. Your function has a lot more impact on the nature of your job than does the type of company in which you work.

You can have a wide variety of business jobs in the energy sector:

- Asset development
- Corporate finance
- Quantitative analytics, risk management
- Trading, energy marketing
- Investment analysis
- Consulting
- Business development
- Banking
- Strategy and planning
- Economics and policy analysis

Different companies can have widely varying names by which they refer to these roles. For example, "marketing" in one company involves advertising and product promotion, whereas "marketing" in another can mean commodities trading. Similarly, "business development" can be more akin to sales in one company, or synonymous with strategic planning in another.

What Type of Company?

Job functions and company types intersect in numerous ways – for example, you can do corporate finance in a large oil company or with a small fuel cell manufacturer, or choose between asset development and trading within a given utility. See Figures 2.1 and 2.2 for a complete list of the job functions available at each type of company. Below, we have summarized the characteristics of each of the major energy sector employer types:

Visit Vault at **www.vault.com** for insider company profiles, expert advice, career message boards, expert resume reviews, the Vault Job Board and more.

V/\ULT CAREER LIBRARY 59

Oil companies

Oil companies engage in exploration and production of oil ("upstream" activities), oil transportation and refining ("midstream"), and petroleum product wholesale and retail distribution ("downstream"). The largest companies, known as the "majors," are vertically integrated, with business operations along the entire spectrum from exploration to gas stations. Smaller oil companies, known as "independents," are often exclusively involved in exploration and production. Upstream is considered the glamorous place to be, where all the big decisions are made. Upstream jobs also involve heavy international work, with many employees sent off to new postings around the world every 3 years or so. We should also note that E&P businesses are fairly similar in nature among oil companies and companies mining other natural resources like uranium or coal – moving among these types of firms during a career can be a logical path.

The majors are known for excellent rotational training programs, and a fair number of people take advantage of those programs and then jump over to independents for good salaries. Oil companies pay well in general, but jobs are not necessarily as stable as one might think. When oil prices drop, company operating profits are dramatically impacted, and layoffs are fairly common. American oil jobs are overwhelmingly concentrated in Houston. International hot spots include London, Calgary, and the Middle East.

Some oil companies focus exclusively on midstream and downstream activities. They operate refineries to distill crude oil into its many commercially useful petroleum derivatives, like gasoline, jet fuel, solvents, and asphalt. Refineries are, in theory, built to last 40 years, but some have been around for as long as 80 years. That means that new refineries are rarely built, and the refinery business is mostly about managing the razor-thin margins between purchased crude oil inputs and revenues from refined product outputs.

Oil services companies

Oil services companies provide a very wide range of outsourced operational support to oil companies, such as owning and renting out oil rigs, conducting seismic testing, and transporting equipment. The fortunes of these companies follow the price of oil: when oil is expensive, oil companies drill a lot and make a lot of money, so business volume and revenue increase for their oil services contractors. Working for an oil services company probably means working in Texas or internationally, and can feel very much like working for an oil company, given the similarity in issues and activities.

Pipeline operators

Pipeline operators own and manage tens of thousands of miles of petroleum products and natural gas pipelines. Many of them also operate oil intake terminals, engage in commodities trading and energy marketing, and own natural gas storage facilities or petroleum refineries as well. Unlike the major oil companies, pipeline operation companies are not household names – nonetheless, the largest ones take in several billion in annual revenue, comparable to the scale of a medium-sized oil company.

Utilities

Utilities are, by definition, located all over the country..everyone has to get their electricity and gas from somewhere, of course. However, as a result of massive consolidation among utility holding companies, the corporate offices for your local utility may not necessarily be that local. There are presently about 50 investor-owned utilities in the country, but industry insiders predict that in a few years mergers may leave us with as few as 10. The "graying" of the utility industry is a well-documented trend; 60% of current utility employees are expected to retire by 2015 – meaning there's lots of opportunity today for young job seekers.

"Utility" is actually a loose term that we use to succinctly refer to gas utilities and all types of power generation companies: investor-owned utilities, government-owned utilities, municipal power companies, rural electric co-ops, and independent power producers (IPPs) or non-utility generators (NUGs). Utilities differ greatly in terms of their lines of business: some have sold off most of their generation assets and are primarily distribution companies with power lines as their primary assets; others may own large amounts of regulated power plants, and may also own non-utility generators or individual independent power plants. As the electricity market fell apart starting in 2001, most IPPs sold off their assets piecemeal to large utility holding companies or financial institutions.

Transmission grid operators

Transmission grid operators, known as Independent System Operators (ISO) or Regional Transmission Operators (RTO), provide a power generation

Visit Vault at **www.vault.com** for insider company profiles, expert advice, career message boards, expert resume reviews, the Vault Job Board and more.

VAULT CAREER LIBRARY   61

dispatch function to a regional electricity market. They don't own the transmission lines, but coordinate how much power is generated when and where, such that supply and demand are equal at every moment. This is an extremely complex process, and necessitates the analytical skills of electrical engineers and other generally quantitative and analytical operations staff.

Equipment manufacturers

Equipment manufacturers make turbines, boilers, compressors, pollution control devices, well drilling and pipeline construction equipment, software control systems, pumps, and industrial batteries. Many of them also provide engineering services and construction/installation of their equipment. The major gas turbine manufacturers, for example, also offer engineering, procurement and construction of entire power plants. Oil-related equipment makers are often characterized as "oil services" firms (above). The equipment manufacturers in the energy industry are not particularly concentrated in one geographic area, though of course many of the oil business-oriented ones have major offices in Texas.

Investment funds

Investment funds are a diverse bunch: mutual funds, private equity funds, and hedge funds. As a whole, the investment fund world is fairly concentrated in Boston, New York and San Francisco, but there are small funds dotted all over the country as well.

Mutual funds hire stock analysts primarily out of MBA programs to track, value, and recommend stocks in a particular sector (e.g. energy, natural resources, consumer products) to the fund managers. However, there are a lot of other finance-related positions inside these massive firms where undergrads are sought after as well.

The number of hedge funds in the U.S. has been growing at a phenomenal rate in the past few years, but they are still notoriously difficult places to get jobs. Hedge funds often hire people out of investment banking analyst programs. They tend to not hire people out of the mutual fund world, given that their valuation approach is so different, their investing horizon is so much shorter, and their orientation many times is towards short-selling as well as buying stocks. While some hedge funds may focus exclusively on energy, most are generalist and opportunistic with respect to their target sectors.

Private equity funds invest money in private (i.e. not publicly traded) companies, often also obtaining operating influence through a seat on the portfolio company's board of directors. As a result, an analyst's work at a private equity fund is vastly different from that at a mutual fund or hedge fund. You are not following the stock market or incorporating market perception issues into your valuations and recommendations; instead, you are taking a hard look at specific operating issues, identifying concrete areas where the portfolio company can lower costs or enhance revenue. A few private equity firms specialize in energy investing, and many more do occasional deals in the energy space as part of a broader technology or manufacturing focus. Private equity firms hire just a few people straight out of college or MBA programs, and many others from the ranks of investment banking alumni.

Banks

Banks are primarily involved in lending money to companies, but they also have their own trading operations, private wealth management, and investment analysis groups. Commercial and investment banks arrange for loans to energy companies, as well as syndicate loans (i.e. find other people to lend the money) for them. Investment banks manage IPOs and mergers and acquisitions (M&A) activities as well. The banking world is overwhelmingly centered in New York (and London), with some smaller branches in Chicago and San Francisco.

Consulting firms

Consulting firms offer rich opportunities for those interested in the energy industry. Consulting on business issues (rather than information technology or technical, scientific issues) is done at three types of firms: management consultancies, risk consulting groups, and economic consulting shops. Consulting firms are often interested in hiring people with good functional skills rather than requiring specific industry expertise, and provide a broad exposure to energy sector business issues, as well as good training. Business consulting firm offices are located in most major cities, but much of the energy sector staff may be located in Houston, Washington D.C., and New York.

Visit Vault at **www.vault.com** for insider company profiles, expert advice, career message boards, expert resume reviews, the Vault Job Board and more.

V/\ULT CAREER LIBRARY   **63**

Nonprofit groups

Nonprofit groups are tax-exempt corporations (pursuant to IRS code 501(c)3) engaged in issue advocacy or public interest research. Advocacy groups may focus on developing grassroots support for public policy changes, publicizing public interest issues or problems through direct actions, or working to influence politicians to enact or change legislation. Most of the energy-related advocacy groups focus on environmental topics, though some also cover corporate financial responsibility and investor protection issues. Think tanks are public policy research institutes, staffed mainly by PhDs who generate research and opinion papers to inform the public, policy-makers and media on current issues. Interestingly, the think tank is primarily a U.S. phenomenon, although the concept is slowly catching on in other countries. Some think tanks are independent and nonpartisan, whereas some take on an explicit advocacy role. Nonprofits are funded by individual donations and grants from foundations, and accordingly a substantial portion of their staffs are dedicated to fundraising. Most energy nonprofits are based in Washington, D.C., where they have access to the federal political process, but many of them have small regional offices or grassroots workers spread out across the country.

Government agencies

Government agencies at the federal and state levels regulate the energy markets and define public energy and environmental policy. Federal agencies are mostly located in Washington D.C., and each state has staff in the state capital. Jobs can include policy analysis, research project management, or management of subcontractors. The energy agencies tend to hire people with environmental or engineering backgrounds, and are lately following a policy of hiring people with general business and management education and experience.

Energy services firms

Energy services firms help companies (in any sector) reduce their energy costs. Working for an energy services firm is similar in many respects to consulting-except that you go much further down the path of implementation. Typically, an energy services firm first conducts an energy audit to understand where a company spends money on energy: electricity, heat, and industrial processes. Then, the firm actually implements energy-saving measures "inside the fence" of the client company. This can involve investments and activities such as putting lightbulbs on motion sensors, upgrading the HVAC (heating, ventilation, air conditioning) system, negotiating better rates with the utility suppliers, or developing a cogeneration power plant adjacent to the factory. Often, the energy services firm receives payment for these services in the form of a share in the net energy cost savings to the client. These firms are located across the country, with a few of the largest clustered in Boston.

Figure 2.1: Employer Types by Job Function

Job Function	Possible Employer Types
Asset Development	Utility; Oil Company; Pipeline Operator; Energy Services Firm
Corporate Finance	Utility; Pipeline Operator; Oil Company; Equipment Manufacturer
Quantitative Analytics, Risk Management	Utility; Oil Company; Transmission Grid Operator; Pipeline Operator; Investment Fund; Bank
Trading, Energy Marketing	Utility; Oil Company; Pipeline Operator; Investment Fund; Bank
Investment Analysis	Investment Fund; Bank
Consulting	Consulting Firm; Oil Services Company
Business Development	Equipment Manufacturer; Utility; Oil Services Company; Pipeline Operator; Energy Services Firm
Banking	Bank
Strategy and Planning	Utility; Oil Company; Pipeline Operator; Oil Services Company; Equipment Manufacturer
Economic and Policy Analysis	Government Agency; Nonprofit Group; Consulting Firm

Visit Vault at **www.vault.com** for insider company profiles, expert advice, career message boards, expert resume reviews, the Vault Job Board and more.

VAULT CAREER LIBRARY   **65**

Figure 2.2: Job Function by Employer Type

Employer Type	Possible Job Functions
Oil Company	Asset Development; Corporate Finance; Quantitative Analytics; Risk Management; Trading; Energy Marketing; Strategy and Planning
Oil Services Company	Consulting; Business Development; Strategy and Planning
Pipeline Operator	Asset Development; Corporate Finance; Trading, Energy Marketing; Business Development; Strategy and Planning
Utility	Asset Development; Corporate Finance; Quantitative Analytics; Risk Management; Trading, Energy Marketing; Business Development; Strategy and Planning
Transmission Grid Operator	Quantitative Analytics, Risk Management
Equipment Manufacturer	Corporate Finance; Business Development; Strategy and Planning
Investment Fund	Investment Analysis; Trading, Energy Marketing; Quantitative Analytics, Risk Management
Bank	Banking; Quantitative Analytics; Risk Management; Trading; Energy Marketing; Investment Analysis
Consulting Firm	Consulting; Economic and Policy Analysis
Nonprofit Group	Economic and Policy Analysis
Government Agency	Economic and Policy Analysis
Energy Services Firm	Asset Development; Business Development

Startups

As you might expect, because the energy world is so technology-intensive, it is full of early-stage firms trying to bring something new to the marketplace. Because startups often haven't earned any revenue yet (or have earned revenue but no profits), pressure is high and hours long. At the same time, the opportunity to impact the business strategy and participate in executing the strategy are relatively high, even for less experienced people. People who really want to "make a difference" tend to find small entrepreneurial companies appealing. Much of the industry's innovation happens in these smaller, newer firms, so they can be a logical place to begin if you really want to be on the cutting edge.

Before you get caught up in the glamour of a particular startup opportunity, make sure you ask three key questions:

1. **How much and what kind of funding does the company have?** You may certainly want the stability of working for a company whose founders are self-funding operations. However, if the company has substantial outside funding from a reputable venture capital firm, then that's one more vote of confidence in the company's prospects. There are tons of entrepreneurs out there in the energy world – one filtering tactic is to follow the smart money and let it do part of the selection for you.

2. **What value does the company's service or product bring to the energy market?** If you can't understand what value there is in what the company is offering, then there's some chance that the market as a whole won't either. While people talk about how working for failed startups is a great learning experience, working for a successful one is usually an even better experience! So, try to join a winning team. Furthermore, if you don't believe the value proposition, those long hours will be hard to stomach.

3. **What is the commercialization pace of the company?** There are plenty of fuel cell companies that have been trundling along with prototype after prototype for years, with expectations of sustainable profitability still years in the future. In other words, not every "startup" is on a path towards launching new products or services in the near future. Make sure you understand the company's strategy and timeline, and how that impacts the job responsibilities and office environment.

Visit Vault at **www.vault.com** for insider company profiles, expert advice, career message boards, expert resume reviews, the Vault Job Board and more.

V∧ULT CAREER LIBRARY   **67**

Energy Hiring Basics

CHAPTER 5

Who Gets Hired?

As in other technology-intensive sectors, the energy sector is populated by a disproportionate number of people with technical degrees, i.e. BS, MS, or PhD in engineering, hard sciences, and math. Whether it's true or not, traditional energy company employers often feel that success in a job correlates to having a certain degree. This pickiness about your undergraduate major or master's degree field gets even stronger during economic downturns, when companies act more conservatively and have more bargaining power in terms of new hires.

In many energy jobs, the prevalence of people with technical pedigrees is somewhat a function of self-selection: individuals interested enough in the energy sector to make it their career were usually also interested enough in related topics to focus on them academically. On top of that, the prevalence of technical people is also self-reinforcing; in other words, engineers like to hire other engineers. There is also arguably an element of reality underpinning the preference for people with certain academic backgrounds – engineers communicate best with other engineers, and have proven in school that they can learn the ins and outs of a complex subject area.

This tendency is most characteristic of hiring preferences among oil companies, oil services firms, refineries, pipelines, grid operators, equipment manufacturers, energy services companies, and utilities. These firms want to hire people who have their heads around how their technologies work – people who can master the jargon quickly, and who can fit into their culture. Even for their MBA hires, these companies often look for technical undergraduate degrees or pre-MBA work in energy or another technical field.

However, there are certainly many people with liberal arts backgrounds doing great work at these types of companies. A non-technical degree does not in any way shut you out of any energy sector career path; it simply makes you slightly more unusual in the eyes of some interviewers. If you can craft a compelling story about why you are passionate about and deeply understand the energy world, your degree becomes far less relevant. In addition, if you are applying for a finance, economics or accounting job with a degree in those fields, you are also less subject to scrutiny about your knowledge of geology, electrical engineering, or chemistry. Once you have a couple years of

Visit Vault at **www.vault.com** for insider company profiles, expert advice, career message boards, expert resume reviews, the Vault Job Board and more.

V\ULT CAREER LIBRARY **69**

experience in the industry, that serves as a degree equivalent and you will have established your credibility.

Many of the service jobs in energy are interested in simply hiring smart people who demonstrate an ability to learn a new industry quickly. Energy consulting, banking, and investing jobs often screen for nothing different than their counterparts in other industries. Similarly, the newer, alternative energy companies are often heavily filled with people who studied liberal arts, economics, and government in college. These companies are progressive in terms of their business strategies, and usually this comes across in their approach to hiring as well. In addition, nonprofits typically first look for passion and commitment to advocacy work before they look for technical background.

Apart from academic background, traditional energy employers are also keenly interested in people who have a strong connection to the geographic region in which the company is located. These companies like to hire for the long term, so will often grill out-of-state candidates about why they would want to move to, for example, Houston or Atlanta. This can mean that, for a Houston oil company position, an MBA from Rice is a more attractive candidate than one from Wharton.

In fact, the energy sector offers particularly rich opportunities for students from second tier undergraduate and graduate schools. Energy companies know that their industry is not typically considered as hot and glamorous as some other industries, and they can therefore often be skeptical about recruiting from name-brand undergraduate and graduate schools. The bottom line is that energy, as an industry, is simply less hung up on name-brand schools than some other industries, i.e. consulting, law and banking.

Moreover, during the past few years of our sluggish economy, many traditional energy companies tightened their recruiting budgets and reduced focus on first-tier schools – at the same time as service companies like consulting and banking firms reacted to a slow economy by canceling recruiting at second-tier schools and concentrating on only a limited set of top schools. Of course, those in the know are well aware that the energy sector is one of the most intellectually challenging, influential arenas in which to work! If you want to work in the sector, you can certainly seek out the energy employers, regardless of whether they visit your campus or target people from your alma mater.

In general, the best time to jump into the energy sector is right out of undergraduate or graduate (MA/MS, MBA or PhD) school. Like most employers, energy companies expect less in the way of industry experience from people who have just graduated, so it's a good time to get your foot in the door of a new field. Lateral hires of people a few years out of college or post-MBA are relatively rare, unless you have some specific industry background or functional experience a company needs. For example, a pipeline company might realistically hire someone with a couple years of general banking experience into a corporate finance role, but would be very unlikely to hire someone with a couple years of, say, real estate experience into that same role – so if you had just graduated and never spent those couple of years in real estate, you'd have a better shot at the job.

This reluctance to hire laterally from other industries is far less common in the services sector (consulting, banking, investing, nonprofits). These employers are more interested in functional knowledge and pure brainpower, rather than a track record in one particular industry or another (though they have their own intransigence about hiring people laterally from other functional areas, i.e. it's awfully hard to get into consulting or banking if you don't do so your first year out of school). As a result, these jobs are an excellent way to get into the energy sector, and offer lots of options down the road – in other words, for example, it's relatively easy to go from an energy consulting role into a corporate job at other energy firms.

One caveat for those who move from one firm to another to position themselves for a future job: traditional energy employers like stability. If you have a lot of different jobs on your resume, you should make sure to have a good story to explain the necessity of your job-hopping, and why you are long-term play for the company (whether you truly are or not). This is true when interviewing with any firm, but large, traditional energy firms are certainly more sensitive to the issue.

Overcoming the Experience Paradox

We've all heard it many times: "industry experience required." But if all the jobs in the industry specify that, where are you ever supposed to get your first industry experience?

The answer is: you don't. People who have a painless experience getting their first job in the energy industry typically have managed to find some type of "experience" to put on their resume before they ever get a job. With good knowledge about the industry, they are able to articulate why they want a job in energy, demonstrate passion for it, and communicate effectively with the industry insiders who are interviewing them.

Visit Vault at **www.vault.com** for insider company profiles, expert advice,
career message boards, expert resume reviews, the Vault Job Board and more.

V/\ULT CAREER LIBRARY   **71**

What might your first experience be if it's not a real job?

- Participating in the energy club at school
- Doing an academic research project in the field
- Taking a one-off class about some aspect of the industry
- Creating an "internship" by doing some free work for a professor or company
- Or simply doing your reading and being able to relate a compelling story about your commitment to the sector.

Sincere interest in a job can come only from in-depth knowledge about the job. Thus, use your cover letter and your interview to demonstrate how much you already know about energy, and as a result how driven you are to work in the field.

Having gathered some good knowledge about energy, you can focus on getting your first job. Some types of companies are more interested than others in hiring people into their first energy sector job. In general, consulting firms, government agencies, investment banks, and the upstream departments of oil companies offer formalized internships and rotational training programs which are specifically geared to people fresh out of undergraduate or graduate school, or lateral hires with no industry experience. In contrast, many of the other energy sector employers lack formalized programs for people new to energy, meaning that jobs there are just harder to find – not that they are nonexistent.

In addition to looking for companies that are explicitly open to people with no prior industry experience, you should attempt to leverage the background you do have. Given your academic and professional background to date, there will definitely be jobs that require more or less of a strained explanation about how they link to your past. For example, if you have a real estate or legal background, you can make a compelling case for why you want to work in energy asset development. Similarly, with an economics degree but no formal energy exposure, you can still be a convincing candidate for market analysis and strategy jobs. Or, if you've been working in finance or engineering, you can be very appealing to energy trading desks, despite not having working in energy previously.

Internships

One way to gain knowledge about and demonstrate interest in the energy business is to pursue an internship in your field of choice, whether it is financial analysis in an oil company, consulting, investment analysis, economic analysis in a utility company, etc. A summer internship during college or business school will expose you to work very similar to that of a BA-level or MBA-level new hire in the company you intern for. In addition, working on the job for a few months is by far the best way to find out if you like the work, the culture, and the energy business in general. If your schedule is flexible, consider looking for a spring or fall internship, or working part-time for a local company during school.

In most cases, as an intern you would be handed some piece of segmentable work that can be carved out for you and completed in the course of a summer – perhaps a special market analysis that nobody has had time to dig into, assisting on a big transaction that needs extra bodies, reviewing and refining models or analysis that can stand some focused attention. Some companies who haven't thought through their need for an intern may dole out less meaningful work that doesn't build skills or industry knowledge terribly well. Either way, as an intern you should work hard to make the most of your experience. Build your industry network by meeting as many company employees as possible; set up one-on-one meetings with employees in departments other than your own to understand the business more broadly; ask tons of questions and keep notes about what you do and learn so you can translate the internship into a compelling full-time job interview.

Most firms don't have formalized internship programs, but rather accept one-off interns in various departments for the purposes of both cost-efficient labor and recruiting. These firms generally won't come looking for you to join them for a summer – you need to get on the phone, use your network, make cold calls and write letters to identify or create a position for yourself. Unless you have a specific connection you can leverage to get an "in," big corporations are better targets than small companies, as they have more capacity to take on interns.

That said, a few types of energy sector employers do have formalized intern programs: consulting firms, large investment management firms, and investment banks. These companies hire fleets of interns out of colleges and business schools, and use the summer job to evaluate them for full-time job offers. These internships are almost always obtained through structured, on-campus recruiting programs, and will include valuable skill training and performance reviews through the course of a summer. Some government agencies, nonprofits, and many of the larger oil companies also take on a handful of interns each summer.

Tangible vs. Intangible Work

One of the big decisions to make in entering the energy sector is whether to work directly with energy products, or whether to work in an advisory or supporting capacity with companies who themselves work directly with energy products. In other words, do you want to work for the pipeline operator, or for the bank that lends to the pipeline operator? Should you work for the turbine manufacturer or the fund that invests in the turbine manufacturer? What about joining the utility versus the government agency that regulates the utility? And how about the choice between the oil company and the nonprofit that advocates for policy changes in the oil company?

Looking at Figure 2.3, inside the circle, your primary job responsibility is analyzing, managing, or coordinating a flow of BTUs, electrons, or barrels of oil that your employer directly controls. Outside the circle, your primary responsibility is analyzing, managing and coordinating financial and information services, advice, and rules to facilitate those BTUs, electrons and barrels of oil making their way through the economy. Simply put, the companies inside our circle control BTUs directly, and those outside the circle do not.

The choice of whether to work with things versus ideas should be based on your personality and interests. Inside-the-circle jobs are conceptually hands-on – you may not literally roll up your sleeves, but you are directly involved in operating issues. People who work for the power generators, oil producers, and grid and pipeline operators can point to examples of their work while driving down the highway. Even if you are a lowly financial analyst inside a big oil company, the construction of a new LNG facility, for example, feels legitimately like the fruit of your and your team's labors.

On the other side of the divide, in the world of professional and financial services, regulatory oversight, and equipment supply, we find a far fewer number of jobs in total. In fact, there are twice as many inside-the-circle business jobs out there as there are outside-the-circle positions. However, these service and supply jobs are often very high value: steeper learning and experience curve, faster pace, and higher pay. In this world you are accountable for the creation of a study regarding the economics of oil production, rather than oil production itself; you work to solve problems on paper for energy companies, but are a step farther removed from energy production and distribution itself.

Many people working in professional and financial services companies and equipment supply companies define themselves first in terms of their job function and second in terms of their participation in the energy industry. For example, an energy investment banker will be hired as a generalist, and think of herself as a banker first, and an energy person second. Similarly, the content of work in energy consulting and strategic planning in an oil company may be rather similar, but the consultant thinks of himself as such and may more readily move out of energy and into another industry.

The service and supply jobs – the jobs outside the realm of direct control of BTUs – attract an overwhelming portion of the MBAs entering the energy sector. In part, this reflects traditional stigmas about what is and is not prestigious; but that reputation is linked in turn to who has traditionally offered higher salaries. The high-pressure consulting, banking, and investing jobs in particular are attractive to people who don't require a stable, low-stress lifestyle, and can trade off longer work hours for what can often be a faster career path, more challenge and more money. In addition, these firms often have somewhat more varied locations across the country – entering the energy industry in consulting, for example, doesn't force you to move to Texas the way entering the oil business usually does. Services jobs in particular not only carry a great deal of prestige, but they keep your options open – you can work in a consulting firm or bank and retain many options as to where to go next, whether it's to somewhere else in the energy sector, or to another industry altogether.

Visit Vault at **www.vault.com** for insider company profiles, expert advice, career message boards, expert resume reviews, the Vault Job Board and more.

VAULT CAREER LIBRARY

75

Figure 2.3: Energy Sector Participants

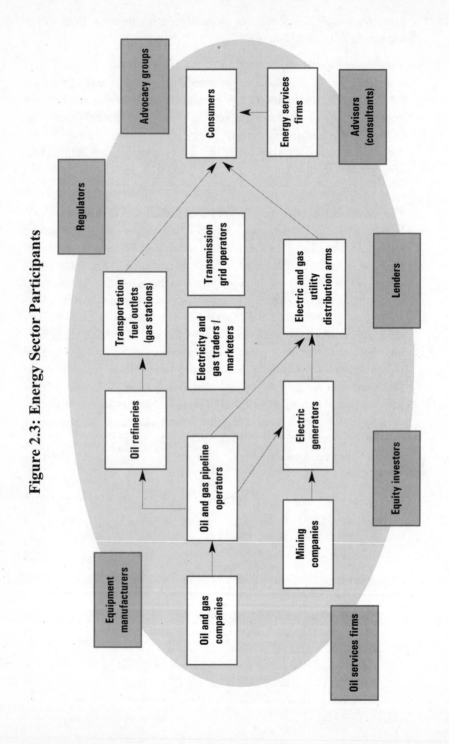

"Good Guys" vs. "Bad Guys"

Ethical issues and values are played out dramatically in the policies and investment decisions of energy companies. Unlike most other industries, corporate activities in the energy sector can have a potentially large adverse effect on the health of each of us and the health of our environment. Unlike in many other industries, choosing among energy sector job offers may force many candidates to consider whether the company's activities are in line with their personal values.

The difficulty is that, of course nobody wants to work for a bad company, and at the same time no company does only bad stuff; so, the "good guys" are sometimes hard to tell from the "bad guys." The reality is that most jobs – even in the high impact energy sector – are dominated on a daily basis by work that most people would likely feel rather neutral about. It is only the outliers, a very few jobs in a very few organizations, where one could truly spend all day doing unquestionable good or bad in the world. Most individuals working for the world's most notorious wrongdoer corporations are actually trying to do the right thing. Don't, for example, imagine that oil companies are full of people who want their children to develop asthma as a result of air pollution.

Fundamentally, people are happy in jobs where they can be themselves and express their personal values. If you are a naturally talkative person, chances are you won't be happy programming in Visual Basic all day. Likewise, if you are a passionate environmentalist, working in a high-volume polluting energy company could either be frustrating, dangerous for your job security, or even perhaps rewarding as you work as an agent for change from the inside. Ultimately, you need to identify what is important to you in order to find the right fit with an employer and a job function.

People talk about energy companies engaging in "greenwashing": publicizing pro-environmental actions that are trivial or actually self-serving, when in reality they are focused on profits. This practice means it's difficult to tell PR from sincerity – but what does sincerity mean in the context of corporations? The notion of fiduciary responsibility means that, by definition, a corporation makes choices based on contribution to the bottom line and benefits to shareholders (which means they are looking out for you — if you have any money in the stock market via a 401k, pension, or mutual fund, or if you want employment at the company). So, if a company advocates tightening emissions restrictions on coal-fired power plants because it holds a lot of natural gas power plants and would realize a relative

Visit Vault at **www.vault.com** for insider company profiles, expert advice, career message boards, expert resume reviews, the Vault Job Board and more.

V∧ULT CAREER LIBRARY

77

benefit, does that mean the impact of its lobbying against a dirty fuel is less valuable? If companies pursue windpower development because it's currently one of the only ways to make good money in asset development, does that mean that these renewable energy facilities are less clean?

Oil companies are certainly the most derided "bad guys" in the energy world. Yet, for all their history of ignoring the health, human rights and environmental impacts of their activities, there are hardly any oil companies left that aren't at least now moving in the right direction. The pace in which they are moving, of course, varies substantially. Shareholder activism in the past few years has been successful in affecting oil corporation policies: companies have committed to including a carbon price per ton when evaluating new projects, or reporting on their renewables investing and greenhouse gas reduction activities. Conoco-Phillips made the recent notable move of dropping out of the Arctic Power lobbying group that promotes opening the coastal plain of the Arctic National Wildlife Refuge for oil and gas drilling.

Most large private oil companies have embraced environmental stewardship....	...but a few have been less enthusiastic about change.
• Royal Dutch/Shell (UK/Netherlands) has achieved major voluntary greenhouse emissions reductions, and has significant hydrogen and solar businesses.	• ExxonMobil (U.S.) actively lobbies against climate change action, including funding ultra-conservative and anti-environmental organizations.
• BP Amoco (UK) was the first big oil company to acknowledge the reality of climate change (in 1997). It has already voluntarily reduced GHG emissions to 10% below 1990 levels (at zero net cost) through internal global cap-and-trade system, and invests actively in solar energy.	• Unocal (U.S.) has a particularly poor record of oil spills and human rights abuses abroad. It has been widely criticized for harmful behavior in Indonesia, including toxic chemical releases and severe ground pollution.
• ChevronTexaco (U.S.) was the first big U.S.-based oil company to follow in BP's footsteps in acknowledging the reality of climate change (in 2000); Chevron actively funded climate change skeptics, but the merged company has moderated such tactics.	• Many of the largest oil companies in the world are government monopolies in countries like Saudi Arabia, Mexico, Venezuela, China, Nigeria, Kuwait, and Brazil. These entities operate without the benefit of shareholder pressures, and have generally lagged behind the private companies environmentally.

Despite the oft-cited facts above, determining which oil companies have better and worse records overall is difficult. BP may have been a real leader on climate change to date, but the company is still interested in drilling in the Arctic National Wildlife Refuge despite vehement protest from environmental groups. Similarly, Shell has a better-than-most record on pollution issues, but has had major accounting scandals, and was infamously implicated in human rights abuses in Nigeria not too long ago. Exxon, in turn, is generally viewed as having horrific environmental policies, but ironically has the highest accounting standards among its peers.

As long as we need oil to live our lives, companies will be able to profit from oil drilling. And, as long as companies can profit from oil drilling, companies will drill for oil. We can't ask the oil companies to stop providing us the fuel to drive our cars, run our factories, heat our homes, and fly our airplanes. All we can realistically ask is that companies invest meaningfully in future technologies that will eventually reduce our dependence on a polluting, depleting fossil fuel resource.

So, if you are considering working in the energy sector, but uncomfortable with the ethical implications of doing so, what choices do you have?

- Work for an oil company to help oil exploration and production occur as efficiently as possible, so more money can be diverted toward other activities like environmental compliance or alternative fuel technology development.

- Send a message by asking questions about the company's position on the issues you care about during your interviews. If they are threatened or annoyed by your questions and concerns, they won't hire you. If you don't hear answers that make you comfortable, you don't have to work there.

- Make your views known inside the company once you start working there. BP, after all, started its "Beyond Petroleum" campaign as a direct result of employee dissatisfaction with working for big oil in an era of escalating public concern for the environment.

- Work for a company that you believe is, on balance, doing good according to your value system – a renewable generation developer, for example.

- Work in the public sector to create new policies and incentives that shape corporate behavior. Corporations, like people, respond to incentives – if somehow oil production becomes less profitable, companies will respond accordingly and produce less oil.

Visit Vault at www.vault.com for insider company profiles, expert advice, career message boards, expert resume reviews, the Vault Job Board and more.

VAULT CAREER LIBRARY

79

VAULT

THE MOST TRUSTED NAME IN CAREER INFORMATION

Vault guides and employer profiles have been published since 1997 and are the premier source of insider information on careers.

Each year, Vault surveys and interviews thousands of employees to give readers the inside scoop on industries and specific employers to help them get the jobs they want.

VAULT

The Interview

CHAPTER 6

Interview Styles

Interviews for most business jobs, regardless of sector, are fairly informal and conversational. Engaging in a chatty interview full of "why should I hire you?" types of open-ended questions is very common, particularly in more traditional companies. That said, interview styles can vary widely from one particular organization to another, so you must always prepare carefully for an interview: practice answering the questions you might be asked, and making a persuasive case for why you should be hired. Specific questions you should always prepare for are:

- Tell me about yourself and your background.
- Walk me through your resume.
- Describe a typical project or problem from your current or previous job.
- Why do you want to work here?
- How do you know you want to work in the energy industry?

Be natural and speak about what you've done in a way that's relevant to the job and conveys what you are proud of. Fit matters enormously to most employers, and the only way you can both assess if personalities are a good fit is if you just be yourself.

Your interviewer may also ask you content-related questions about energy problems, particularly if you are not fresh out of college and have relevant job experience to tap into. Typically, these are not at all confrontational, but simply an effective way to further identify which person would fit best in the job. If you are asked to explain how to price a natural gas swap, for example, it is usually not a quiz on which you need to actually calculate the answer. Rather, the employer is interested in seeing how well you can explain the process, how comfortable you are thinking through it out loud, how naturally you use the job-specific lingo, and what your demeanor is when discussing something complex.

Asking good questions in return is just as important as answering the prospective employer's questions well. This is your chance to further convey your passion for the subject matter of the job at hand. Ask the interviewer what he or she is working on at the moment. Actively listen to the response – it's amazing how smart one can sound simply by playing back or restating what someone told you, to show that you listened and understood. If you

Visit Vault at **www.vault.com** for insider company profiles, expert advice, career message boards, expert resume reviews, the Vault Job Board and more.

V∧ULT CAREER LIBRARY 81

have familiarity with the specifics of what the interviewer is working on, feel free to offer up your own ideas in the form of questions: "I did a project once that was somewhat similar. Have you looked at the problem in this way…?" Another trick is to ask if you can see an example of their work output (Excel model, PowerPoint presentation, printed report, memo). Seeing the actual physical product of the work that goes on in the office can quickly give you a rich understanding of the job.

While most business job interviews are fairly conversational, there are a few job categories that involve specific types of structured interview questions. The following services roles have unique interview styles that require advance study and thorough preparation:

- **Consulting:** Case interviews are fairly standard among consulting firms. You may get business cases on energy, or on other industries as well. Some employers use cases that are similar to their actual work, and some use generic cases written by third parties. Energy cases could be questions such as thinking through a retail gas station strategy or evaluating the growth prospects for diesel cars in the U.S. Cases conducted by associates are usually very by-the-book, requiring a by-the-book, framework-driven response. In contrast, cases with partners are typically more conversational, and a venue where your creativity is more appreciated.

- **Banking:** Above all, bankers want to hear that your first and only passion is banking. They are notoriously aggressive and confrontational with interviews, so you need to have a perfectly watertight "story" that links together all of your past experiences with your current interest in working for their specific firm. Common questions in banking interviews include: explain the responsibilities of an investment banker, who else are you interviewing with, explain each of your academic and professional decisions, explain how to value a company, summarize market activity in the past few days. MBA candidates in particular can expect tough questions on accounting, securities pricing, valuation, and financial theory.

- **Investing:** Investment management firms will expect you to know a handful of stocks in their industry focus areas well enough to make a persuasive and comprehensive pitch. In addition, they will expect you to have your own active investment portfolio, be able to talk in detail about relevant coursework, and chat about current trends in market indices and policy issues affecting prices. For example, make sure you walk in knowing what happened to oil prices this week, and where electricity prices have been this season in the major markets.

Sample Interview Question #1: Valuing a Power Plant

"How would you structure the analysis for a power plant investment?"

This open-ended, tell-me-what-you-know type of question is something you could run into in a corporate finance, investment banking, investment management, or even a strategy job interview. You might hear this question in many forms: "Do you know how to value a power plant?" "Do you think the sale of Company X's assets was overpriced?" "Does Generation Company Y present a good investment opportunity?" A finance professional needs to answer these questions routinely – when a company or investor is involved in developing a new generation facility, bidding on plants being sold at auction, pricing assets for divestiture, or valuing an asset-owning business. Principals of valuation apply to all types of assets, so if you can talk about power plant investing in an interview, you can also comment intelligently on investment issues for pipelines, oil rigs, and the like.

You may not be asked to go in-depth with specific numbers in your interview. Nonetheless, in order to answer this question successfully, you do need to understand a fair bit about what's involved in building and operating a generic power plant. We can look at specific cost figures for a gas-fired combined cycle power plant, which is the most common type of new power plant built in the U.S. A good rule of thumb to keep in mind is that new capacity costs about $600 per kW. Due to significant economies of scale, a 250MW plant might cost more than $150 million (250x1000x600), and a 1000MW plant might come in at something under $600 million (1000x1000x600). Turbine costs are a significant percentage of overall cost, and are thus carefully negotiated with vendor companies, many of which offer a bundled "EPC" (engineering, procurement, construction) contract for the facility construction and turbines.

Visit Vault at **www.vault.com** for insider company profiles, expert advice, career message boards, expert resume reviews, the Vault Job Board and more.

VAULT CAREER LIBRARY   **83**

Major Construction Cost Components for a 500 MW Gas Power Plant	
• Facility construction	50 %
• Turbine purchase	25 %
• Capitalized interest during construction	7 %
• Transmission interconnection and system upgrades	5 %
• Mobilization costs	4 %
• Fuel and electricity for testing	
• Labor	
• Spare parts inventory purchase	
• Gas pipeline connection	3%
• Land purchase	2 %
• General and administrative costs	2 %
• Costs from development phase	2 %
Total cost	**$300 m**

Operating a power plant profitably is all about running it during hours when electricity prices are higher than fuel prices, and hoping that hourly operating margin (revenue minus variable cost) adds up to enough over the hours you run during the year to cover all of your fixed costs-and also contribute something to paying back the capital invested to build the plant in the first place. In our example, we assume the following representative figures:

- Electricity prices average $40/MWh during the hours our plant is dispatched, which is 80% of the 8760 hours in the year (40 x 500 x 8760 x 80% = $140m).

- Fuel (gas) averages $3 / MMBtu during those same hours, and our plant converts fuel to electricity at a very efficient heat rate of 6500 Btu/kWh. (3 x 6500 x 500MW x 8760hours x 80% = $68m).

- Other variable costs add up to $0.50/MWh (0.5 x 500 x 8760 x 80% = $2m). These are primarily fees paid to the host community for use and discharge of water to cool the plant, which can be some 100,000 gallons per day.

- Fixed costs are $100/kW-year (100 x 500 x 1000 = $50m). Nearly half of these costs are for labor and administration. Well over a third of fixed expenditures are typically for regular annual maintenance (a plant is often shut down for a week for its annual overhaul) and for funding the "major" maintenance reserve (major portions of the plant must periodically be replaced due to wear-and-tear).

Sample Pro Forma Income Statement for a 500MW Gas Power Plant (millions)	
Revenue	$140
Costs	—
Variable costs	—
Fuel	$68
Other	$2
— Water supply	—
— Chemicals for water treatment	—
— Water discharge	—
— Purchased power for startup	—
— SO2 emission allowances	—
— NOx emission allowances	—
Total variable costs	$70
Fixed costs	—
Labor and administration	—
Regular maintenance	—
Major maintenance	—
Insurance	—
Property tax	—
Total fixed costs	$50
Total costs	$120
Net operating profit	$20

So, how do you actually answer the question in the interview? You probably don't walk through the cost figures laid out above, which are here primarily for your background knowledge. Remember, the interviewer is likely not trying to test your ability to manipulate numbers in your head without a calculator – rather, s/he is trying to see if you understand valuation on a conceptual level. Thus, you need to frame your analysis but not actually execute it. Generally, you will want to communicate your grasp of three basic aspects of valuation:

1. Discounted cash flow valuation methodology. You need to demonstrate that you know how to do a basic, textbook DCF valuation: Take revenues, subtract variable costs, fixed costs, taxes, change in net working capital and capital expenditures to yield free cash flow for each year; calculate a Net Present Value (discount the FCF stream back to the investment year using the firm's cost of capital valuation, add the initial capital cost); conduct sensitivity analysis on the major assumptions.

2. Back-of-the-envelope valuation methodology. Perhaps more importantly, you also need to demonstrate that you can assess a project's value intuitively, without any fancy Excel-based analysis. With just four numbers – expected energy price, fuel price, heat rate and capacity factor – you can easily comment on the financial viability of a plant:

   • Expected energy price minus variable cost (expected fuel price x heat rate) is your expected hourly operating margin.

   • Multiply that by the expected hours run (hours in a year x capacity factor), and you have the annual operating margin.

   • Ask your interviewer what the plant's fixed costs are. If they are less than your annual gross operating margin, then you know the plant can be expected to at least break even.

On an even more qualitative level, one can reasonably comment on the financial viability of a prospective new gas plant with even just one number – the heat rate:

   • The lower a plant's heat rate is, the lower its variable (fuel) costs are.

   • Power plants are generally dispatched in ascending order of bid prices (variable costs): the lower the bid, the more the plant will be called to run.

   • However, market prices are set by the highest bid, so if your costs are substantially lower than the most expensive dispatched plant, you stand to make money by virtue of receiving a higher price for your output than it costs you to produce it.

- Thus, if the plant in question has a heat rate substantially lower than the plants which are generally the most expensive ones dispatched (and you can observe this by looking at a graph of the market's "stack" of available plants), you know by definition that the plant will make money.

3. "Pro forma" numbers vs. actual results. Asset valuations can be done with varying levels of detail and with more or fewer simplifying assumptions, depending on the stage of the project and the purpose of the analysis. If you do a rough valuation for a proposed new development, you may have 20 line items in a 1 megabyte Excel file. On the other hand, if you prepare a valuation to support a $100 million loan for an existing plant, you may have 200 line items in a 3 megabyte model. You will want to demonstrate awareness that assumptions like "annual average hourly electricity price" and "annual average fuel price" do not reflect the complexity of actual operations. In practice, a power plant may earn revenue from ancillary services, payments for capacity availability, and additional income from power marketer transactions. Much of its output sales (and its fuel purchases) may occur through long-term contracts rather than the daily commodity market. The heat rate is not a constant, but varies with on/off cycling of the plant and with the temperature outside. As forecasters are fond of saying, any model of the future is almost certainly wrong.

Sample Interview Question #2: Strategizing About Climate Change

"So what do you think should be done about global warming?"

For an environmentally-oriented or public advocacy job, this question could be the crux of your interview. In an oil or electricity company, you might be asked a question like this in the guise of a fit interview or seemingly innocuous hallway chit-chat. But, you need to be aware that even idle conversations are part of your interview and the impression you make on the prospective employer.

First, be aware of your audience. Interviews are a time to be honest and not misrepresent yourself, but at the same time they are not a platform for any strong, politically-motivated views you may have. You want to be honest so

Visit Vault at **www.vault.com** for insider company profiles, expert advice, career message boards, expert resume reviews, the Vault Job Board and more.

V/\ULT CAREER LIBRARY   **87**

that you don't end up getting hired and working for a company filled with people who radically disagree with your own beliefs – but at the same time you do want to get a job.

A balanced, carefully reasoned argument about the climate change issue should always be acceptable, no matter where the company's incentives lie. For example, a coal company should be willing to hear you say that the preponderance of evidence is toward warming and that you believe their industry is sufficiently innovative to develop technology solutions to the problem. Similarly, a wind generation company should be willing to hear you point out that wind cannot be the entire answer, because windpower plants cannot be built in sufficient quantities to fully offset fossil-fired generation, and they are only cost-competitive in very large installations that local residents often oppose.

Secondly, make sure you provide a structured answer. Energy companies are notorious for having less formal interviews, where the interviewer isn't prepared with a formal case question and may not have specific criteria against which to evaluate you. But an apparently casual question nonetheless deserves an organized response.

In this case, a good answer could proceed with an exchange such as the following:

Candidate: "Well, the way I see it, if we want to reduce carbon dioxide emissions as a society, we have three alternatives:

- We can switch to low-carbon fuels

- We can extract carbon dioxide from the emissions streams of fossil fuels

- We can find ways to use less fuel period, and thus produce less carbon dioxide

Switching to low-carbon fuels involves building more renewables, like wind and solar. Some people advocate nuclear as carbon-free option too. And then the long-term vision of a low-carbon fuel solution would of course be the proverbial hydrogen economy-fuel cells running off hydrogen produced by renewable energy. In the short term, developing the natural gas sector is another relatively low-carbon approach, since natural gas produces a lot less carbon than coal does."

Interviewer:	"Hmm. The problem with focusing on natural gas is that it becomes a crutch fuel and we get stuck with that interim solution permanently."
Candidate:	"True – that is a possible scenario. Which would necessitate exploring our second option simultaneously: extracting carbon from the waste streams of the fossil fuels we use. Carbon sequestration technology is advancing and getting more cost effective, and to be realistic, next generation clean coal technology can be useful in reducing emissions too. Ultimately, though, I think most of our near-term opportunities for emissions reduction fall in the third category of simply using less fuel."
Interviewer:	"That's certainly a popular opinion. How easy do you think it really is to simply 'use less' as you say?"
Candidate:	"Well, it is disturbing that with energy use correlated to economic growth, the much-needed economic growth expected in developing countries will drive a massive increase in total energy consumption. However, I for one am a firm believer that energy efficiency often pays for itself and thus is not a difficult sell. I'm very hopeful about continuing improvements in internal combustion engine and turbine efficiency, efficient natural gas fuel cells, the expanded use of cogeneration, and recovering otherwise wasted landfill and flare gas as fuel. The telecommuting trend may even make a palpable dent in the total amount of driving."
Interviewer:	"So you think that improving gas mileage in our SUVs can offset the industrialization of the third world?"
Candidate:	"It's a tough problem, as you point out. I'll be the first to acknowledge that. The solution probably has to come simultaneously from many of these changes: switching to low-carbon fuels, extracting carbon from emissions, and reducing consumption. No one action could be sufficient to realize the amount of reductions people are talking about."

What makes this answer a good one? This candidate organized thoughts into three categories, and answered the question according to the pyramid

Visit Vault at **www.vault.com** for insider company profiles, expert advice,
career message boards, expert resume reviews, the Vault Job Board and more.

VAULT CAREER LIBRARY   **89**

principle of communication: summarize your argument at a high level first, then provide the detailed supporting examples and logic. The candidate was able to use the interviewer's interjections and questions to guide the discussion back to the original three points and stay on message. When challenged by the interviewer in a possibly argumentative fashion, this candidate responded well – she maintained a balance between conceding the interviewer's point, yet also standing firm to her own well-reasoned opinion. To close out the discussion, the candidate made sure to reiterate the original main point of her answer.

Now, for many people looking for their first job in the energy sector, the level of detail in the model answer above may not be realistic. What's most important, though, is not demonstrating that you can rattle off twelve solutions to an environmental problem that has confounded our society for years, but demonstrating that you understand the concepts and broad categories of possible actions. If you are asked a question that seems to require a lot of detailed content knowledge that you don't feel you have, then say so. Fair interviewers should generally have no problem with your saying, "I don't feel like I have all the facts on this issue to make a judgment. Could you walk me through some of the details of the issue first?"

Sample Interview Question #3: Commercializing a New Product

"We have a new fuel cell design ready for manufacturing, but as you know fuel cells aren't yet in widespread use. Tell me how you would think about taking our product to market."

A fuel cell company employee might simply ask you a question like this conversationally, or a consulting firm might ask it of you in a more formalized case format. Either way, a good answer starts with stating what you know, asking questions to gather more information, and doing a lot of active listening. You are not going to generate a comprehensive commercialization strategy in 20 minutes, and the interviewer most likely has their own, well-thought-out answer to the question at hand – let the interviewer guide you to the specific issue that they are interested in discussing with you.

To address this interview question, you need a fair amount of background knowledge about fuel cells. Having done your homework for the interview, you are likely well versed in the topic. If not, you need to gather the information by asking a series of specific questions of the interviewer:

- The company's fuel cell uses a proton exchange membrane (PEM) design, considered to be the most promising and widely applicable. The company has launched a 5 kW model priced at about $20,000.

- PEM fuel cells are generally expected to be used to power cars and provide back-up power for commercial facilities. Like most fuel cells, they generate electricity from natural gas at a somewhat higher cost per kWh than the average U.S. retail price, so they are not considered to be greatly appealing as primary generation sources.

- Unlike larger types of fuel cells, PEMs operate at the relatively low temperature of 80° C, so their water byproduct is hot, but not hot enough to form steam to run a turbine in a cogeneration configuration. As a result, their efficiency is lower and cost of energy output higher.

- Currently, a number of fuel cell manufacturers have products on the market, but very few backup power or primary generation installations have been completed to date, as adoption has been slow. The automotive market has not materialized at all.

For any question about product commercialization or technology marketing, you can use the "4 P's" framework as a way to organize your thinking and remind you what questions to ask:

Topic	Key questions
Product	What is the product? What are its physical characteristics? How long does it last?
Price	Does the sales price make sense, relative to substitute products? Does the sales price make sense, given the product's manufacturing cost? How much does it cost to use on an ongoing basis?
Promotion	What promotional channels can we use to communicate our product's benefits to the target market? Which aspects of the product's value should we emphasize?
Place	Who is our target market? Who are the likely early adopters? How should the product be distributed? Where will buyers obtain it?

Visit Vault at **www.vault.com** for insider company profiles, expert advice, career message boards, expert resume reviews, the Vault Job Board and more.

VAULT CAREER LIBRARY   91

In this case, one of the key issues for the company's sales success with their new fuel cell product is clearly finding a large, receptive target market. Electricity is a commodity, and thus electricity generators compete on a price basis. So, you want to outline for your interviewer the major price-related issues you wish to explore:

1. The high cost of electricity produced by fuel cells is a major stumbling block to their widespread adoption.

    a. Is there a market that doesn't mind paying higher-than-retail prices for power from fuel cells?

    b. Is there a market in which the cost of electricity from our fuel cell is competitive with or lower than the prevailing retail price?

2. In the case of this PEM fuel cell, there is a significant inefficiency in wasting the thermal energy output, which further contributes to a high cost of power. Is there a way to capture that wasted energy?

The interviewer tells you that your first question is interesting: there is a small market of environmentally-oriented individuals and businesses who might on principle pay higher prices for fuel cell-generated electricity. However, that market is finite and cannot provide the sales volume needed to recover all of the product's R&D costs and sustain the company. The more promising price-insensitive market is among commercial institutions that value having a backup power source that can fill in for an unreliable grid: hospitals, sensitive financial operations, police and government offices are some examples. However, our company's product is too small for most of those applications; large, stationary, high-temperature fuel cells are dominating the backup power market.

More interesting is the question of where the price of electricity from the grid is actually more expensive than what can be generated by our company's product. The interviewer points out that while the average U.S. retail price is lower than the product's operating cost, some areas of the U.S. have much higher-than-average prices. Unfortunately, in those areas (Hawaii, New England), the cost of natural gas is also very high, which means that the fuel cell doesn't have any advantage. However, the interviewer reveals, in Japan electricity prices are double U.S. prices, and natural gas is plentiful and not exorbitantly priced. Cheap natural gas fuel in Japan means our fuel cell can produce cheap power there, which will compare favorably to the very high market price of power otherwise available there.

At this point, it becomes clear to you that the company has been looking at a strategy of ramping up sales of its new product in Japan. Seize onto that as the assumed logical commercialization path, and continue by bringing up your third point: what if we could capture the waste heat and somehow increase overall efficiency of the product to make it even more appealing to the Japanese market?

One creative idea, you propose, is capturing the waste heat to use for household hot water or space heating. The fuel cell's electrical output runs household appliances, while its thermal output displaces the need to run the hot water heater or furnace.

Your interviewer asks you what questions you would need answered to validate the feasibility of your idea. Again, you can reach for the "4 P's" framework to remind you what key issues need to be addressed for any new product introduction. The obvious "fatal flaw" questions would include:

- Is the hot water output from the fuel cell hot enough for household hot water or space heating? Is there enough of it to make an appreciable dent in a household heating bill?

- Is the kW output of our product appropriate, given the electrical load in a typical Japanese home?

- Do Japanese utilities offer net metering, whereby the excess power generated can be sold back to the grid so that the fuel cell doesn't have to run inefficiently at part load when the household doesn't require the full output?

- How would the effective cost of power and heat from a fuel cell compare to a typical Japanese household's current alternatives?

- Can we integrate the fuel cell and hot water systems together effectively, from an engineering perspective? Would the installation cost be prohibitive?

- Do most Japanese homes have enough space to install our size fuel cell? Do most homes in Japan already have natural gas hookups?

- Are a typical household's electrical and hot water/heating demands correlated in time? If not, is the thermal output from the fuel cell storable?

Visit Vault at **www.vault.com** for insider company profiles, expert advice, career message boards, expert resume reviews, the Vault Job Board and more.

V\ULT CAREER LIBRARY   **93**

Sample Interview Question #4: Oil Exploration Risk Analysis

"Our company is considering development of some oil deposits off the coast of western Africa. Through our agreement with the local government, we have access to three separate oil fields, but we can only pick one. How do we value the fields and decide which one to pursue?"

Large oil companies outsource just about every element of their E&P processes: they rent many of their oil rigs, contract geologists for seismic testing, and use third-party shippers. The primary activity they retain in-house that can offer them a competitive advantage is investment decision-making. These companies commonly make multi-billion dollar bets, so they had better be very good at investment decision analysis. This sample interview question represents a very typical analysis that a finance person in an oil company would undertake, as well as the analysis that the company's lenders and consulting advisors might conduct.

Background

Start by gathering background information on the situation, which is only very generally described in the question. Your interviewer may offer up some of this information, but you will likely need to figure out the right questions to ask to solicit this information. A good approach is to begin your response by stating, "I'd like to begin by asking a series of questions to gather more information on the situation so that I can determine the best solution methodology."

- Big Oil Company purchased production rights from the government of the African country. Its agreement provides for royalty payments to the local government for any oil that is actually extracted.

- Oil fields have very high internal pressure, and much of their contents would gush out very quickly if it could. However, wells are drilled and gathering pipelines are built with an economical width and capacity that ends up constraining the extraction rate. As a result, oil flows out of the well at a more constant rate over a longer period of time (See Figure 2.4). When you evaluate a given oil field, you must take the binding constraint into account to determine how much oil would actually flow, and when – oil extracted today is usually worth more than oil extracted in the future, due to the time value of money.

- Big Oil Company conducted seismic tests at each field location in order to verify the presence of oil. However, such testing can produce false positives. Based on the conditions at each site, engineers can calculate a probability of geological success for each oil field – in other words, a probability that the identified oil below the ocean floor in fact exists.

- Big Oil Company's development committee has authorized the funding of the development of just one of the African offshore fields. Similar to most exploration and production companies, Big Oil prefers to evaluate prospective projects in terms of both Net Present Value and Investment Efficiency (NPV divided by capital investment).

Approach

Now that you better understand the context for the interviewer's question, you can explain your approach to solving the problem. It is always best to announce beforehand the method you intend to use, so you appear organized and your thought process clear. A statement such as the following will score you big points: "I would recommend that we calculate an NPV for each of the three fields, taking into account potential oil volume extracted, operating costs of the well, and capital costs to develop the field. We will need to multiply potential revenues and operating costs by the probability of success in each case. I think it would be interesting to look not only at NPV and IE, but also at the magnitude and probability of potential losses in each case, since NPV just reflects the expected outcome but not the distribution of possible outcomes."

Data

In order to complete the calculations you propose, you need a fair amount of input assumptions. In some cases, the interviewer (when asked the correct question) will produce a prepared set of data assumptions for you to work with. Alternatively, you can simply make your own educated assumptions in order to do the calculations.

- Production for any of the three fields would start in 2008, after well and trunkline construction is completed. It will then take two years to reach peak production volume. Decline from peak to zero production also lasts two years. (See Figure 2.5)

- Big Oil Company uses a standard flat price of oil in its valuation models: $20/barrel. (Oil companies are fanatically secretive about their host government royalty agreements and their oil market price forecasts; this price assumption incorporates both the complex royalty agreement and the company's proprietary market price outlook.)

Visit Vault at www.vault.com for insider company profiles, expert advice, career message boards, expert resume reviews, the Vault Job Board and more.

VAULT CAREER LIBRARY

95

- Big Oil Company's discount rate is 10% (i.e., the accounting factor the company chooses to compare future cash flows to today's dollars). Normally, you would need to discount each year's cash flow individually by that year's appropriate discount factor. To save you calculation time, your interviewer provides you with a composite discount factor that should be applied to each field's total revenue over time.
- We assume no taxes.

Data provided by the interviewer						
	Peak volume (M bbl/year)	Years at peak	Probability of success	Composite disc. factor	Initial capital cost	Lifetime operating costs
Field 1	120	3	50%	50%	$800 M	$800 M
Field 2	200	6	25%	50%	$1,200 M	$1,200 M
Field 3	60	9	80%	40%	$1,500 M	$1,500 M

Calculation Details

Now that you have the requisite data, you can proceed with calculating the NPV for each field:

1. Calculate the potential lifetime oil production volume from each of the three fields, which is the area under the curves in Figure 2.5.

2. Multiply each field's volume by $20/barrel to yield each field's potential lifetime undiscounted revenue.

3. Multiply by the composite discount factor to yield the potential lifetime discounted revenue.

4. Finally, multiply by each field's probability of success to yield the expected lifetime discounted revenue.

5. For operating costs, take the lifetime cost and multiply by the composite discount factor. Then, make sure you also multiply by the probability of success – remember, if the well fails, then not only do you not have any revenue, you don't incur the associated operating costs either!

6. Subtract the capital and expected operating costs from expected revenue to arrive at the expected NPV for each field

7. Divide NPV by initial capital cost to yield the investment efficiency (IE) index.

You will end up with the following results:

Results of interviewee's calculations				
	Potential oil volume	Potential revenue (undisc.)	Potential revenue (disc.)	Expected revenue (disc.)
Field 1	540 bbl	$10,800 M	$5,400 M	$2,700 M
Field 2	1500 bbl	$30,000 M	$15,000 M	$3,750 M
Field 3	630 bbl	$12,600 M	$5,040 M	$4,032 M

			2	3	4	
	Potential operating costs (undisc.)	Potential operating costs (disc.)	Expected operating costs (disc.)	Initial capital costs	Expected NPV 1 − (2+3)	IE (4 ÷ 3)
Field 1	$1,500 M	$750 M	$375 M	$800 M	$1,525 M	1.9
Field 2	$2,800 M	$1,400 M	$350 M	$1,200 M	$2,200 M	1.8
Field 3	$1,500 M	$600 M	$480 M	$1,500 M	$2,052 M	1.4

The key to getting these calculations right is to remember to incorporate probability of success. When we value many assets, we simply use revenue and operating cost assumptions without any further adjustment. In the case of oil wells (and any other investment that has a possibility of totally failing), it is crucial to understand the probability of ever getting those projected revenues and incurring those projected costs. To talk about expected value, we must multiply everything in the future by the probability of success. Paying the capital costs, in contrast, is not a function of whether the oil ends up being extractable or not, so we don't adjust those.

Conclusions

Don't stop doing the calculations, as the most important part of a good answer is interpreting and summarizing the results. There are many observations you can now make about the choice before Big Oil Company:

- All three fields are attractive investment opportunities, as they have positive NPVs.

Visit Vault at www.vault.com for insider company profiles, expert advice, career message boards, expert resume reviews, the Vault Job Board and more.

VAULT CAREER LIBRARY   97

- Field 2 has the highest NPV. In other words, it is expected to bring in the most profit to Big Oil Company.

- Field 1 has the highest IE. In other words, it is expected to bring in the most profit to Big Oil Company per dollar invested.

- Field 3 has an NPV very close to that of Field 2, but a much lower IE. Big Oil Company would have to put up a lot more capital to generate the same bottom-line impact as with Field 2.

- We can also calculate expected losses for each of the fields: probability of failure multiplied by capital cost. For example, Field 1 has a 50% probability of losing its $800 million capital investment – an expected loss of $400 million. Similarly, Field 2 has a $900 million expected loss, and Field 3 $300 million.

- While Field 2 yields the highest NPV and a high IE, it also has the lowest probability of success, at 25%, and the highest expected loss.

  — Depending on Big Oil Company's risk appetite, the company may want to consider developing Field 3, which has a high NPV, yet has the lowest expected loss. However, if Field 3 does fail, Big Oil Company would not recover any of its $1.5 billion investment.

  — Risk aversion might also steer us to prefer Field 1, which has a fairly low expected loss ($400 million), the lowest potential loss ($800 million), the highest investment efficiency index (1.9), and a positive NPV.

Depending on Big Oil Company's risk metrics, the "right answer" could conceivably be any one of the three available oil fields. What is important is not that you pick one as your answer, but that you walk through the pros and cons of each choice, and demonstrate that you can think about value along a number of dimensions.

Additionally, to take your response to a valuation question like this from good to great, you can point out other possible project risks and propose some creative alternatives:

- How much political risk is there? Is there a chance the local government could pull out of its agreement with Big Oil Company midway through the process, leaving the investment stranded and unrecoverable?

- Does the proposed development impact any local populations in a way that might incite protests? Social turmoil is bad PR for Big Oil, and causes financial losses from operational disruptions – not to mention the injustice ofnegative impacts on the host country's inhabitants.

- Are there design modifications we can make to ensure that, whichever field is developed, the operation has minimal environmental impact? Have we factored in the expected costs of environmental mitigation into our project valuation?

- Is there additional testing we can do to raise our confidence in the presence of a large oil deposit in any of the fields? In particular, if we had more confidence in Field 2's success, it could emerge as the clear winner for development.

- Perhaps Big Oil Company could create a partnership or otherwise raise the additional capital needed to invest in more than one field. Investing in two fields would increase the overall probability of success with at least one field. For example, if we were to invest in both Fields 1 and 2, and we assumed their probabilities of success are independent, our chances of success would increase to 63% (1 – 50%*75%), which is higher than either of the two fields alone.

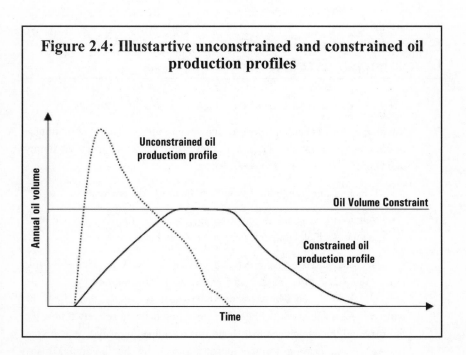

Figure 2.4: Illustartive unconstrained and constrained oil production profiles

Visit Vault at **www.vault.com** for insider company profiles, expert advice, career message boards, expert resume reviews, the Vault Job Board and more.

VAULT CAREER LIBRARY

99

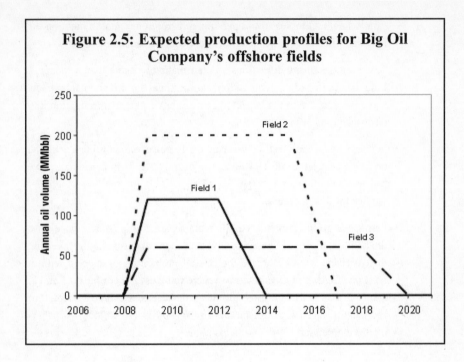

Figure 2.5: Expected production profiles for Big Oil Company's offshore fields

Sample Interview Question #5: Pitching a Stock

"Tell me about an energy stock that you think is a good investment."

Pitching a stock is central to equity analyst interviews. However, in almost any energy sector job interview, it is conceivable that someone might simply ask you:

- "Which of our competitors should we be most concerned about?"

- "You said you're interested in energy because it's dynamic and fast-growing – which companies are you referring to?"

In such cases, you can give a detailed answer similar to the one below, just minus the stock price assessment at the end. In addition, you should generally read up on M&A trends and regulatory rulings that affect the particular field you are targeting. Demonstrating that you actively follow the industry and the companies that shape it is a compelling way to prove your passion – and passion for the industry is one of the top things many companies seek.

For either an equity analyst or other interview, you should have a story like the following one on deck for 2-3 stocks or companies that truly interest you. Remember that, particularly when talking to hedge funds, discussing an over-valued or poorly-performing company can be a compelling vehicle to show off your analytical ability.

Be sure to think through ahead of time how you will communicate your pitch. In the following answer, for example, the interviewee uses an intuitive structure that would be easy for an interviewer to follow: quick summary of who the company is and why it's compelling, market size, competitive positioning, and current/future financial performance.

"A company in the energy sector that I have followed with particular interest is ABC Batteries, which manufactures batteries that power cell phones, and is in the process of commercializing a battery designed to power hybrid-electric cars. Currently, Japanese battery manufacturers dominate the hybrid-electric battery market, but U.S.-based ABC is poised for explosive growth as it enters this new market in the United States.

First of all, ABC's new target market is growing. Sales of hybrid-electric cars are likely to increase dramatically in the next few years:

- **Demand for hybrid-electric cars is growing:** Oil prices recently topped $50 per barrel, which is a 20-year high. The operating cost savings between hybrids and gasoline-powered cars is directly correlated to oil prices: as oil prices rise, high fuel economy cars offer greater savings to consumers and become even more appealing. Importantly, consumer preference for high fuel economy cars is driven by expectations of future high oil prices, in addition to actual high oil prices. So, even if oil prices rationalize over the short term, long-term concerns about mid-East instability and declining world oil reserves should support strong consumer demand for hybrids.

- **Hybrid-electrics are now competitively priced:** Hybrids tend to be priced a few thousand dollars more than their standard counterparts. However, the lifetime cost of these cars is actually lower than that of standard gasoline-powered cars when gas prices exceed $1.75 per gallon.

- **More hybrid-electric suppliers are coming to market:** Hybrids are a proven commercial success, thanks to Toyota's recent mass-market release the Prius, which had the fastest-growing sales of

Visit Vault at **www.vault.com** for insider company profiles, expert advice, career message boards, expert resume reviews, the Vault Job Board and more.

V\ULT CAREER LIBRARY **101**

any car the year it was launched. Low sales for the Honda Insight suggested that the market would be small, but Toyota has demonstrated otherwise. The 'Big Three' U.S. auto makers have focused on designing SUVs for the past several years, but with all of the public backlash against gas-guzzling SUVs and the advent of more stringent fuel economy standards (particularly in California), they are all introducing hybrids.

Secondly, ABC is well-positioned to be successful in the hybrid-electric car market:

- **ABC's product is the standard:** ABC licensed the technology for its nickel metal hydride battery product from Energy Conversion Devices, as did all of the Japanese hybrid-electric battery manufacturers. Rechargeable NiMH batteries last twice as long as traditional lead-acid batteries, and are lighter weight, which makes for a lighter car and thus higher fuel efficiency. All hybrid-electrics on the market today use NiMH batteries. Over the long term, we expect to see hydrogen-fueled fuel cell cars come to market, but such technology is far from commercially viable.

- **ABC has sufficient production capacity:** ABC just completed expansion of one of its cell phone battery plants to include capacity to produce 100,000 NiMH car battery modules per year. The company also has the option to convert some of its cell phone battery production capacity to accommodate car battery production if needed.

- **ABC has no U.S. competition:** Sanyo and Panasonic currently have a duopoly on the hybrid-electric car battery market. However, U.S. automakers have previously stated their preference to use U.S. parts suppliers. ABC has a first-mover advantage in the U.S. – as soon as it signs a contract with Ford, GM or Daimler-Chrysler, it will have a lock on the U.S. hybrid-electric car battery market.

- **ABC's product can expect good margins:** The commercial success of a hybrid-electric car is primarily driven by the availability of a reasonably-priced, energy-efficient battery. Therefore, auto manufacturers have leverage to negotiate lower prices for their hybrid-electric batteries, and there is some concern about a 'race to the bottom' among battery-makers. However, with no other U.S. battery makers in the market, ABC should be able to launch its product with good pricing.

- **ABC is in a strong financial position:** ABC's current operations generate enough cash to enable successful commercialization without relying on the debt markets.

Finally, ABC's current stock price does not yet fully reflect the company's substantial growth potential:

- ABC is trading around $10 per share right now, with a P/E ratio of 20x that is consistent with other relatively mature technology manufacturing companies. Assuming that (1) ABC captures half of expected U.S. automakers' hybrid-electric sales going forward, and (2) the market applies a P/E ratio of 40x to reflect the higher-growth profile of ABC's new business, a price as high as $30 would be reasonable.

- In fact, an even higher P/E ratio may be reasonable. The stock prices of oil producers, oil services companies, and energy efficiency businesses react disproportionately to rising oil prices (while, refinery, power generator, and other heavy petroleum consumers' stocks suffer disproportionately). Under a high oil price scenario, I would expect the market to enthusiastically value a company supplying the hybrid-electric car market."

Visit Vault at **www.vault.com** for insider company profiles, expert advice, career message boards, expert resume reviews, the Vault Job Board and more.

VAULT CAREER LIBRARY   103

Use the Internet's
MOST TARGETED
job search tools.

Vault Job Board

Target your search by industry, function, and experience level, and find the job openings that you want.

VaultMatch Resume Database

Vault takes match-making to the next level: post your resume and customize your search by industry, function, experience and more. We'll match job listings with your interests and criteria and e-mail them directly to your inbox.

> the most trusted name in career information™

ON THE JOB

Energy Sector Culture

CHAPTER 7

The energy sector certainly exhibits a distinct culture. In fact, many energy people feel that the culture is so strong that energy sector affiliation trumps functional or organizational affiliation. That means that an energy analyst in a mutual fund would be more culturally similar to a utility strategic planner than she would be to a technology analyst in her own fund. The veracity of this observation really depends on the company – even within a set of the same type of firm, some organizations will have more of an "energy culture" than others.

Office culture is a vastly important determinant of interview success and, later, job satisfaction – so you should be careful to take it into account as you pursue jobs in the energy world. People like to hire other people like themselves. For this reason, one standard interview technique is to mimic the body language of your interviewer to set her at ease (crossed legs, eye contact, speech volume and pace). At the same time, the last thing you want is to get hired into an organization that has a very different personality from yours – you are likely to become frustrated and create friction. Ultimately, there is no substitute for self-awareness as to what type of culture you will thrive in, combined with your own keen due diligence during the job exploration and interview process.

Traditional and Conservative?

A synthesis of how energy businesspeople describe their typical colleague sounds something like this: "a white male electrical engineer who spent time in the military, took some business classes at night, and enjoys fishing and golf." Indeed, the energy sector has a reputation for being traditional, conservative, lacking in diversity, and dominated by dry, technical people. That said, however, we'll tear that stereotype apart by pointing out that in fact many major areas of the energy world reflect great diversity in terms of gender, ethnicity, educational background, and personality. Large oil companies, for example, have a distinct international flavor due to their rotational posting practices that bring employees from all over the world through the home offices in Houston and London. The energy practices of services firms reflect the highly varied profile of MBA consultants, bankers, and investment analysts in general. Startups are populated by the kind of dynamic, aggressive young businesspeople who could just as easily be found in high-tech ventures. Any of the change-oriented organizations – advocacy

Visit Vault at www.vault.com for insider company profiles, expert advice, career message boards, expert resume reviews, the Vault Job Board and more.

VAULT CAREER LIBRARY   107

groups, alternative energy companies, investment funds, independent power generators, consulting firms – are more likely to have employees with a younger average age, liberal arts education, and more progressive orientation.

Like anywhere else in the business world, minority candidates are embraced and even eagerly sought after by firms conscious of workplace diversity. Also like elsewhere in business, ethnic minorities in the U.S. energy sector are still underrepresented, relative to the general population. Energy-sector executives tend to feel that this reflects a mixture of minority under-representation in the academic specialties that lead to careers in energy, self-selection in favor of business sectors or other careers with greater existing diversity, and some residual glass ceiling effects. While minority student job-seekers won't encounter barriers to a career in this sector, they should certainly expect to encounter more white faces than in some of the "newer" industries, such as high tech and biotech.

The one group that has not fared so well to date in the energy sector is women. Because the energy world is dominated by people with technical educations, and women have historically not been well-represented in technical degree programs, they are few and far between at management levels. We have, sadly, heard many stories of talented and hard-working women who are pushed out of positions when they become eligible for management promotions, fired when they have children, asked (quite illegally) about their intention to have children during interviews, alienated from business trips involving hunting or gentlemen's club outings, or simply confronted with skepticism from colleagues and an exhausting upstream swim upon entering the sector.

Like elsewhere in the business world, the proverbial glass ceiling for women is, over time, slowly going away. In the interim, however, if you don't have the appetite for being a female pioneer, there are certainly relatively welcoming portions of the energy sector on which one can focus a career. Consulting firms in general have been relatively progressive in hiring and promoting women, and the energy practices in consulting firms reflect that orientation. Within investment management, mutual fund companies tend to retain women at senior levels (whereas private equity firms and hedge funds are probably the worst culprits across the business landscape in excluding women). Energy policy advocacy groups, government agencies, and alternative energy companies tend to have the highest percentage of women on staff anywhere in the industry. Most banks, private equity firms, hedge funds, oil companies, utilities, pipeline operators, energy services firms, and manufacturers tend to have very few women anywhere but the entry level.

The best way for women to break into male-dominated energy companies is to enter the interview process armed with superior knowledge about the science, technology and economics of the industry. This applies to anyone whose profile makes them a relative outsider to their desired employer's culture – ethnic minorities applying to companies with few familiar faces, people with non-traditional academic backgrounds relative to what is most common in a given company, people who are older or younger or more or less experienced than the other candidates interviewing for a particular job. Ultimately, anyone who wishes to enter the exciting world of energy stands the best chance by positioning themselves credibly as someone who already understands the industry and the specific business problems faced by the interviewer, is passionate about the energy business, is flexible and easy to relate to, and has sincere commitment to the employer's location, work hours, and business mission.

Visit Vault at **www.vault.com** for insider company profiles, expert advice, career message boards, expert resume reviews, the Vault Job Board and more.

V/\ULT CAREER LIBRARY   **109**

Breaking Down the Jobs

Asset Development

Energy is an asset-intensive industry. What people mean by that is that investing in physical structures is the crux of energy company value creation. Power plants, drilling platforms, pipelines, and transmission lines can cost billions of dollars apiece – accordingly, revenues from these assets are applied more to fixed maintenance costs and capital recovery than to variable operating costs. With so much money going into assets, and asset performance driving corporate profitability, many types of energy companies (oil companies, pipeline operators, power generation companies) focus on being careful, smart investors. They often see their competitive advantage as residing in an ability to make better investment decisions than the rest of the playing field, or to protect their investments better than the next guy through risk management.

Working in asset development is one of the most traditionally desirable places to end up in the energy sector. Indeed, many regard "putting steel in the ground" as the most glamorous work in the energy field (jockeying for that distinction with M&A work). As a developer, you are everyone's customer, with lenders, consultants, nonprofits, equipment suppliers all lined up to support the process and get a piece of your business. Inside a company, the development department is often the nucleus of activity and the place that people want to transfer into.

Developing assets involves building new facilities and expanding or retrofitting existing ones. Such tasks involve financial analysis work to justify and plan for a project, and project management work to execute it. Depending on their size and structure, companies may separate the financial analysis role ("analysts") from the project management role ("developers"). Most employers will look for candidates who can master the complex engineering issues that can drive asset value. Backgrounds including engineering education, operational experience, or knowledge of another heavy industry are highly sought after. Analyst roles are sometimes filled with people having less engineering knowledge, but more mastery of complex financial analysis or market economics.

Power generation asset development tends to occur in cycles with lumpy capacity additions: companies overbuild when output (oil, gas, electricity) prices are high, which then depresses prices, resulting in a subsequent period with little development while companies wait for demand to increase back to a level where new supply is needed. Presently, we are in a serious lull in power plant development, after the energy market implosion of 2001. Many independent power producers could not make debt payments on their facilities, and so sold them off to financial institutions (investment banks, private equity funds) who don't mind holding cash flow neutral or negative assets in exchange for flipping them for a lump sum profit in a few years when market fundamentals drive spark spreads higher. Regulated utilities had not invested heavily in generation development since industry restructuring started in earnest in the early 1990s – uncertainty over whether and how utilities would be able to recover any investments they made prevented them from being able to prudently build at all. Now, customers don't want to sign long-term power purchase agreements, and without a PPA, banks won't lend to power projects. Power sector observers anticipate that the next wave of new builds may happen in the 2009 timeframe. In the interim, many large-scale asset developers have found a niche in the ongoing flurry of windpower development.

Another type of asset development to consider is energy services. Energy services firms build and install energy devices and facilities on behalf of industrial and commercial customers seeking to increase efficiency and save money. You could also work in energy efficiency directly for a private, non-energy-related company. Many companies, regardless of industry, have started realizing the amount of cost savings (not to mention environmental stewardship and good PR) available to them by managing their energy supplies and usage. In the energy office for a large paper manufacturer, for example, you could not only work on contracting for electricity and fuel supply at good rates, but also act as developer for inside-the-fence cogeneration facilities, and upgrades/installations of HVAC (heating, ventilation, and air conditioning) systems, boilers, chillers, motors, insulation, energy management software, and lighting controls.

Asset development starting salaries	
New BA-level:	$40,000 – $60,000
New MBA-level:	$90,000 – $130,000

A Day in the Life : Assistant Wind Developer for an Independent Power Company

8:00 a.m.: After waking up in the Holiday Inn in a remote part of Iowa, you get in your rented SUV and drive into town to have a breakfast meeting at the local diner with the mayor. Your development team has optioned a hilltop in the area for developing a windpower facility, and you are now in the process of negotiating a payment in lieu of taxes (PILOT agreement) with the town. It's probably going to end up being a new fire truck, a school playground, and the new access road you will be constructing anyway up to your site. You are flying solo on this meeting, confirming the outline of the agreement with the mayor and putting in some time to continuing to build this crucial relationship.

10:00 a.m.: You drive up to the site to take a look around and call on the farmer who owns the land. The two of you take a walk around the fields together, making note of some exposed bedrock that indicates a spot where it will be too expensive to lay tower foundations. He invites you inside his home for some coffee, and you chat about milk prices. You brought a few photos of wind turbine installations in Europe with cows grazing nonchalantly at the base of the towers, and talk a little about the vast experience with windpower in Europe, where the turbines have been shown to have no negative impact on the underlying farmland or the cattle that call it home.

12:00 p.m.: Back at the hotel, you participate in a conference call with the lead developer and engineers back in the home office. Your team is working up a cost estimate for the transmission system interconnect, in preparation for negotiating the price with the local utility. By law, utilities have to allow independent generators like your company to hook up to the grid, but there's little to prevent them from exacting a high price for doing so. The engineers are preparing some exhibits to counter the anticipated argument that the wind turbines will affect local voltage stability.

1:30 p.m.: Checking your e-mail, you find that one of the turbine manufacturers has responded to your company's request-for-proposal (RFP) for the 20 turbines plus engineering, procurement, and construction services. You'll plug their information into the bid comparison spreadsheet you created when you're back in the office tomorrow. On your voicemail is an "all clear" message from the subcontractor who is conducting the archaeological study of the site

Visit Vault at **www.vault.com** for insider company profiles, expert advice, career message boards, expert resume reviews, the Vault Job Board and more.

V/AULT CAREER LIBRARY **113**

– he was tasked with verifying that there are no remains of historical structures anywhere on the land (which would likely cause the town to deny a building permit).

3:00 p.m.: On your two hour drive to the airport to fly home, you stop at another farmer's home in the next town over. He had contacted your company when he heard that you were planning on building down the road, saying he was interested in showing someone his windy site. You walk around his land together, taking note of flagged trees (permanently bent over at an angle as a result of strong, uni-directional winds). While this is typically a symptom of a good wind resource, there's little conclusion anyone can draw until you install a met data tower and measure actual windspeeds for a year. The wind today feels very gusty, which could be a bad sign – too much turbulence in the local wind pattern causes wear and tear on turbines, resulting in high maintenance costs and shorter equipment life. You tell him you'll bring up the idea of looking more carefully at his site with your team at home.

7:30 p.m.: On the plane ride back to your home base, you do a little reading. Your department is working on formalizing the development process for wind plants in order to save time and money. The idea is to structure the process with a set sequence of activities from low-cost to high-cost and high-level fatal flaw checks to detailed design activities; after each major step in the process, a development oversight committee would decide whether to proceed or not. You read through the draft list to see if there's anything you can add, but it seems very thorough.

Major Steps in Developing a Windpower Generation Plant

- Site selection (based on evidence of wind, proximity to transmission lines, existing road access, receptiveness of local community)

- Detailed wind resource evaluation (review one year of on-site meteorological tower wind speed and direction data, and model turbine output accordingly)

- Land agreements (negotiate royalties, access, facility assignability, indemnification, reclamation provisions with site landowners)

- Environmental review (conduct avian study, wetlands review, check for endangered species, archaeological study, historical review, visual impact study)

- Community agreement (negotiate property taxes)

- Economic modeling (conduct internal review of expected revenues and costs, to present to lenders and equity investors)

- Transmission interconnection studies (work with local utility to identify interconnection point capacity limits and voltage regulation requirements)

- Permitting (obtain local permits for land use and construction, state and federal environmental permits)

- Turbine procurement (evaluate pricing, reliability, and site-specific suitability of turbine alternatives from multiple vendors)

- Sales agreements (negotiate PPAs for output)

- Financing (obtain equity and/or debt to cover capital cost)

- Construction contracting (bid out and negotiate a turn-key contract for excavation, road building, foundation, cabling, tower assembly, turbine installation, interconnection, and commissioning)

- O&M contracting (contract for annual operations and maintenance work, incorporating non-performance penalties)

Corporate Finance

In a utility, pipeline operator, or oil company, the corporate finance group plans and facilitates financing for a company's construction and acquisition activities. Typically, when the corporate finance team becomes involved in a project, a strategy or business development group has determined the parameters of the investment: what should be built, where to build it, how big and with what technology. Corporate finance people identify the means of funding the project and precisely how to structure the transaction:

- Ratio of debt to equity
- Type of debt – private placement, non-recourse, convertible, etc.
- Price of the debt – interest rate, index, fixed vs floating rate

Visit Vault at **www.vault.com** for insider company profiles, expert advice, career message boards, expert resume reviews, the Vault Job Board and more.

VAULT CAREER LIBRARY   **115**

• Terms – life of the loan, tranche structure, repayment covenants, etc.

Additionally, corporate finance validates the revenue and cost assumptions that the strategy people used to justify and receive senior management approval for the project. They prepare a valuation model to be shared with lenders, and market the investment to banks via a detailed offering memorandum and face-to-face presentations.

As with many careers, corporate finance comes in multiple flavors – so make sure you understand the vocabulary of functions specific to each company with which you interview. For companies not heavily involved in developing, acquiring, selling, or funding improvements of physical assets (e.g. refineries, equipment manufacturers), corporate finance groups often incorporate treasury functions: managing working capital, accounts receivable, billing, and financial reporting. They may also handle regulatory compliance, and in particular Sarbanes-Oxley compliance. Large oil companies often have separate Resource Exploration groups that evaluate and purchase rights to drill for oil, while the Corporate Finance group exclusively focuses on financing production fields.

Corporate finance starting salaries	
New BA-level:	$45,000 – $55,000
New MBA-level:	$75,000 – $100,000

	Corporate Finance	Treasury
Oversight	Chief Financial Officer	Controller
Responsibilities	• Manage the buying, selling and financing of physical assets	Controller
Alternate structures	• Includes Risk Management and Treasury functions • Includes a separate Acquisition or Divestiture group	• Part of Accounting or Corporate Finance • Includes a separate regulatory compliance group • Coordinates budgeting and long-term planning

To fulfill all of these responsibilities, corporate finance interfaces with internal strategy, business development, engineering, trading, accounting, and legal groups, as well as outside counsel, joint venture partners, and of course the many banks interested in lending money to the company. In a company particularly active in building, expanding or acquiring assets, the corporate finance job is akin to working in an investment bank.

Corporate finance employees are often generically referred to as "analysts," regardless of level. Formally, however, an "analyst" is usually someone with a B.A., and an "associate" usually has an MBA or a number of years' experience. Depending on the company culture, young associates may actually have more lofty titles such as "principal" or "assistant vice president." Corporate finance jobs don't require a "technical" degree such as engineering, math or hard sciences; however, people with such backgrounds tend to self-select into these jobs, particularly in the oil and gas sector. Generally, employers look for people with experience building complex financial models, keen attention to detail, and a demonstrated interest in their portion of the energy industry.

A Day in the Life: Corporate Finance Associate at a Pipeline Operation Company

8:00 a.m.: Power on your desktop and check e-mail. Most pressing is a message from a prospective lender for the $300 million offshore oil processing platform that your firm intends to build and is now in syndication – a VP of Project Finance at a New York investment bank has some 20 highly specific questions about aspects of the project. You get to work answering them by reviewing the construction contracts, and placing a few calls to the legal department and construction manager.

9:30 a.m.: Sit down with the Corporate Finance Analyst to review her financial model of a planned $200 million oil pipeline development. When you notice that she may have incorrectly set up the construction drawdown schedule, the two of you place a quick call to someone on the Commercial Development team to clarify timing of construction phases. Remind her to incorporate the latest oil price forecast from the Market Assessment team before she circulates the updated valuation to you and your boss.

Visit Vault at **www.vault.com** for insider company profiles, expert advice, career message boards, expert resume reviews, the Vault Job Board and more.

V/\ULT CAREER LIBRARY   **117**

11:00 a.m.: Field a phone call from another prospective lender's Project Finance VP; he has one very specific question about the oil platform's hurricane contingency plan. The facility shutdown criteria are already detailed in the offering memorandum, and you point the banker to a particular page and paragraph number for his answer.

12:00 p.m.: Bring a sandwich up from the company cafeteria to eat at your desk as you peruse an issue of (Structured Finance International) along with this morning's (Financial Times.)

12:30 p.m.: In addition to supporting the current syndication process for the oil platform and developing a model for a future oil pipeline project, you are also finishing up some final transactional details for another pipeline deal that was financed a couple weeks ago. Your firm must document that it has paid in its 30% equity stake before any funds can flow in from lenders; your role is to coordinate with the outside lenders to make sure all the necessary communications happen. A quick sit-down with your firm's financial controller sets the process in motion.

1:30 p.m.: Your entire five-person Corporate Finance team gathers in a conference room to sketch out the contents for the oil platform's bank meeting next month, when you will invite 40 lenders to listen or sit in on a comprehensive presentation of the investment opportunity.

4:00 p.m.: Walking down the hall to grab some coffee from your floor's kitchen, you run into your firm's CFO. You're excited to hear him mention a big joint venture deal on the horizon.

4:30 p.m.: Open up the oil pipeline financial model off the network to review what changes the analyst has made since this morning. You note that she correctly modeled the loan pricing at the current expectation of 300 basis points over LIBOR (the London Inter-Bank Offer Rate). Now it's time to start thinking through the next step – how to build an interest rate swap into the model. Since this type of transaction is less familiar to you, you first skim through an old finance textbook; then you head up a few floors to the trading area where, since the markets have closed, someone can surely give you some insight into the mechanics.

5:30 p.m.: Your calls from first thing this morning have by now all been returned, and you are able to compose a lengthy email response to the interested banker. One of the details about the oil title transfer is still unclear, and you'll have to follow up again with legal in the morning.

7:00 p.m.: After spending some time grinding through the conceptual details of interest rate swap calculations with the analyst, you leave her to execute the change and head home at a decent hour.

Quantitative Analysis and Risk Management

"Quant" jobs can be found with utilities, oil companies, grid operators, pipeline companies, investment banks, and hedge funds. Some consulting firms also do heavy quantitative analysis and risk management work. In addition, for people with an IT-heavy background, development jobs at energy-related software companies can be similar in content to quant jobs inside operating companies.

Depending on the size and structure of the company, you might find separate risk management, regulatory compliance, pricing and structuring, and quantitative support groups, or, all of these functions could be folded into a single team of analytical people who are relied on for any heavy-hitting quantitative analysis needs that arise across the company. Some corporate finance departments also provide the quantitative analysis and risk management functions for their companies.

	Risk Management	Quantitative Analytics
Oversight	• Chief Risk Officer	• Director of Trading
Responsibilities	• Manage and report on market and credit risks arising from trading and operations	• Controller
Alternate structures	• Part of Corporate Finance or Treasury • Includes a separate regulatory compliance group	• Part of Risk Management • Includes a separate pricing and structuring group • Shares employees with the Information Technology group

Visit Vault at **www.vault.com** for insider company profiles, expert advice, career message boards, expert resume reviews, the Vault Job Board and more.

VAULT CAREER LIBRARY   **119**

In smaller companies, one person often has both the CFO and CRO roles. In contrast, most large companies have a distinct CRO, who manages not only financial risk, but regulatory risk as well. Due to the power sector's transactional complexity, compliance with the accounting rules set forth in the 2002 Sarbanes-Oxley Act has become a major function in utilities. You should be aware that, technically, "risk management" can include operational risk management (e.g. insurance, labor relations, supply contracts); however, the most common usage of the term is specifically in reference to financial and regulatory risk.

Quant groups often include a fair number of PhDs in math and physics who can develop complex hedging and asset optimization strategies. Otherwise, these groups are staffed by analytically-oriented people who excel at problem-solving and have great facility with numbers. Good candidates for quantitative analytics and risk management positions often have a background in engineering, programming, and/or "hard-core" finance, which enables them to be good financial model-builders and confident problem-solvers. A mixture of general business and IT skills can be extremely valuable to a Director of Trading or CRO.

Industry experience is not necessarily required for entry-level jobs, particularly if you bring functional experience (Excel modeling, Visual Basic programming, financial engineering). Lateral hires, however, are expected to have a grasp of the complexities of the energy markets.

Quantitative analysis starting salaries	
New BA-level:	$40,000 – $50,000
New MBA-level:	$70,000 – $85,000

A Day in the Life: Analyst in a Quantitative Analytics Group

7:30 a.m.: Arrive at the office at your usual time – work hours are more or less similar to those of the traders, mirroring the operating hours of commodities exchanges. Overnight, you left a position report program running on the server, and now you start the day by reviewing the consolidated results that were automatically dumped into a directory on the office network. To your relief, there are no errors. (Yesterday you spent all morning working with IT to trace the source of an error to a corrupt file in the trading floor system, and then reprocessing the whole desk.) So, you create the daily VaR (Value at Risk, a measure of the financial risk exposure of the company) report and circulate it to the traders, the head of your department, and the Chief Risk Officer.

9:00 a.m.: After taking some time to go through e-mail and enjoy a muffin at your desk, you walk down to the trading floor to meet with a power marketer who wants your help to think through a particularly complex tolling agreement. For most deals, he uses an Excel-based pricing tool you developed last year; however, the proposal on the table has a lot of unique conditions. The two of you work at a white board for an hour or so, drawing illustrative "hockey stick" graphs to conceptually evaluate the contract. One sticking point is that the counterparty proposed a step function for pricing based on power plant output, while your team is more interested in pricing that adjusts more dynamically over time. The power marketer needs specific numerical recommendations from you in two days, so you'll plan to spend time today and tomorrow building an Excel model to support his transaction negotiations.

10:30 a.m.: You attend a department meeting which focuses on the plan to convert most of the Excel tools your group has developed in the past into stand-alone Visual Basic applications (to make them run faster and be easier to maintain). Building tools is exciting, so everyone on your team is clamoring to be involved in this initiative. With your excellent knowledge of VBA (which you use to create fancy macros in Excel), you'll have little problem picking up Visual Basic.

12:00 p.m.: Head over to the on-site company cafeteria to pick up lunch. You see a bunch of folks from IT, with whom you typically interact by phone. You also run into the CFO, who knows you from when you worked with one of his corporate finance analysts on a special project – you ask him how well the balance sheet risk

Visit Vault at **www.vault.com** for insider company profiles, expert advice, career message boards, expert resume reviews, the Vault Job Board and more.

VAULT CAREER LIBRARY   **121**

simulation model you collaborated on is continuing to work for his group.

12:45 p.m.: One of the natural gas futures traders seems to have found a bug in a pricing tool you developed a few months ago. You are constantly working to improve the analytical tools that your team provides to the trading, power marketing, and risk management groups. As people stress-test the tools over time with new and different transaction parameters, they always find small issues, and this time is no different. You manage to locate the bug, correct the problem, and redistribute the tool to the desk team within a couple of hours.

2:00 p.m.: You sit down at your own desk to dig into building the quantitative model that you talked with power marketing earlier in the day. Throughout the afternoon, you periodically walk over to a few other analysts' workstations and trade thoughts with them on the best approach to this tough problem.

3:30 p.m.: Walk over to the kitchenette on your floor to take a short break and fix a cup of coffee. The CRO, who sits on the same floor but on a different hallway, walks in with a couple of the senior accounting managers, talking animatedly about trading limit policies.

5:00 p.m.: On your way out for the day, poke your head into your manager's office to let her know that you're leaving, and will be on track to finish the transaction model for the power marketer by staying late tomorrow night. It's convenient that you can manage your own time in this fashion and make it to a concert tonight with friends.

Trading and Energy Marketing

Any company that produces oil, gas or electricity needs to market the output and hedge its risk exposure through energy marketing and trading functions. That means that jobs in these functions can be had primarily inside electric and gas utilities and oil companies. In addition, investment funds and banks also have substantial energy trading departments through which they gain commodity exposure and profit opportunity.

Energy trading and energy marketing are often loosely referred to together as "trading." They are interrelated activities, and are typically located adjacent to one another inside an office. However, they are very distinct functions that employ different types of people:

- Traders buy and sell standardized commodity contracts to hedge price risk exposure and optimize asset value, and/or to speculate. Traders often specialize by instrument (e.g. options, swaps), rather than by industry.

- Energy marketers and power marketers execute customized bilateral physical transactions to buy and sell (or "market") the oil, gas, or electricity output from their companies' operations. They have specialized regional energy market knowledge.

A single utility company may have no more than 25 people in total across both functions.

A long-standing debate among power traders is whether the company needs to own assets in order to be successful in trading around the output of those assets. In other words, can you make money in electricity swaps if you don't own any power plants? In theory, owning the underlying assets provides a trading group with crucial market information. Most utilities have small trading groups, but they are primarily tasked with asset optimization, rather than pure profit generation. Many investment banks invested in power plant assets after the industry downturn, and are now able to supply their large trading groups with valuable operating knowledge. Hedge funds don't own physical assets, but freely participate in energy commodity trading.

Trading jobs are notoriously good for thrusting a lot of responsibility on new hires right away. You can go into trading right of college with no industry experience. Entry-level traders spend their early months in supporting roles to the more seasoned traders, sitting next to them on the trading floor. Typical tasks include:

- Executing trades
- Gathering and summarizing morning research for the desk team
- Marking positions to market based on daily price changes
- Maintaining daily reports on positions, trading desk profit and loss, settlement activity, trade breaks, and risk metrics
- Ensuring trade and settlement details are correct, and reconciling trade and position breaks with controllers, other traders, brokers, and firm cash management personnel
- Supporting more experienced traders with ad hoc market analysis
- Setting up internal computer systems to accommodate new products, trading accounts and counterparties

Visit Vault at **www.vault.com** for insider company profiles, expert advice, career message boards, expert resume reviews, the Vault Job Board and more.

VAULT CAREER LIBRARY   **123**

Oil, gas and electricity are transported to users across a bottlenecked network of pipelines and transmission lines, with several hubs where standardized trading takes place and from where prices are referenced. While traders are active buying and selling exchange-traded contracts at these hubs, it is the energy marketers who facilitate the scheduling and pricing of the physical flows across the system. Energy marketers (also called "power marketers," "marketers," "wholesale traders" or "energy merchants") engage in highly-structured bilateral buying, selling, and swapping of non-standard products with durations from 1 month to 20 years. The three most typical activities of energy marketers are:

- Physical deliveries: Marketers can act as outsourced energy procurement and energy price risk management specialists for their customers. For example, a power marketer in a utility company might structure a full requirements service contract for some industrial customers or electric co-ops; the power marketer obtains and arranges delivery of electricity, taking on the price risk so that the customer can pay a constant flat price.

- Structured products: In order to maximize revenue for the assets a company owns, its marketers can structure transactions such as tolling agreements, weather hedges, load-following contracts, or complex combinations of exchange-traded futures and options.

- Market making: Some energy marketers are in the business of providing liquidity to the market (for a fee); they take the other side of various transactions, and offset each one with another that preserves a positive margin for their own company.

Power marketers in particular must have extensive industry knowledge and local market understanding. They need to navigate a maze of purchase rules that are specific to each regional power pool. For example, in one pool, regulations require buying power in blocks of 50MWh on a 5-day, 16-hour schedule; in another pool, the standard is 25MWh blocks on a 6x16 schedule. Because of the specific market knowledge required for the job, energy marketers typically come into their jobs from other places in the sector, rather than as new hires directly out of school.

Trading starting salaries	
New BA-level:	$40,000 – $60,000
New MBA-level:	$85,000 – $125,000

Investment Analysis

An investment analysis job is a place where a business graduate can apply a particularly wide range of the topical skills learned in school: valuation, micro and macroeconomics, accounting, and business strategy. In order to assess whether a company is a good investment or not, you need to understand the company's operations and financial results in detail, as well as the industry in which it operates, the strategies of its competitors, and the impact on its operations of macroeconomic changes. Not surprisingly, investment management jobs are popular and highly competitive positions.

Investment analysis jobs can be found within four types of firms:

- **Mutual funds:** The overwhelming number of investment jobs are in mutual funds, which manage more than $7 trillion in customer assets in the U.S. Mutual funds typically invest across all sectors, with industry specialist teams focused on understanding and pitching investments in areas such as energy. Some fund companies, however, do have industry-specific energy funds.

- **Hedge funds:** Somewhat akin to unregulated mutual funds, these funds take larger risks, trade more often, and are free to "short" stocks in order to bet on a bear market. Hedge funds are distinguished amongst one another primarily by their trading strategies (e.g. arbitrage, event-driven, macro, short-selling), and are usually opportunistic rather than industry-specific. The U.S. hedge fund industry is growing extremely rapidly, now managing close to $1 trillion in assets.

- **Investment banks:** Sell-side analysts value public companies to develop stock price targets, and then pitch the stocks to investors in order to generate trading business for the bank.

- **Private equity funds:** Similar to hedge funds in terms of risk profile and overall assets under management, these funds invest in privately-held companies with the aim of improving the portfolio firm's profitability and exiting at a hefty profit through a resale or IPO. In the case of the energy sector, that often means investing in startup energy technology and equipment makers, buying into distressed power plant assets, and investing in oil exploration initiatives through small private firms.

Visit Vault at **www.vault.com** for insider company profiles, expert advice, career message boards, expert resume reviews, the Vault Job Board and more.

VAULT CAREER LIBRARY   **125**

Private equity funds come in a variety of flavors, deserving of some further exposition:

- Venture capital funds (VCs) focus on early stage investment opportunities. VC's often focus on seed stage funding, or late-stage (pre-IPO) funding.

- Growth equity funds, mezzanine funds, and private equity partnerships focus on private companies that are more established, often with significant operating and profit histories. Confusingly, "private equity" can refer both to the whole category of investing in non-public companies, as well as to the non-VC, non-LBO segment of the field.

- Leveraged buyout funds (LBOs) focus on purchasing and optimizing the assets of established operating companies, often with the help of significant leverage, or borrowed capital.

These naming conventions are far from absolute, however. Many private equity partnerships (PEPs) use leverage to invest in large chunks of a company, and thus blur the line with LBOs; similarly, a fund investing in a post-revenue, pre-profitability business might label itself either a VC or a private equity firm.

Where are people investing now in the energy sector? Money flows upstream and downstream along the energy value chain, depending on commodity prices and demand levels. Generally, when fuel prices are high, power generators do poorly, while oil services and oil and gas production are highly profitable, drawing in investment dollars. With sustained high fuel prices, money also flows into demand reduction opportunities: energy efficiency technology companies and alternative fuel startups, for example. In addition to reflecting absolute commodity prices, money shifts around in the system based on volatility of prices; as commodity price volatility increases, hedge funds weight their portfolios more heavily in energy to take advantage of the resulting profit opportunities.

Unlike many of the other company types in which you can get an energy job, investment funds typically value functional skills over industry expertise. In a mutual fund, analysts routinely switch sectors, applying their valuation and business strategy evaluation skills equally well in energy and consumer products or manufacturing. Hedge funds are very thinly staffed, and thus require the flexibility to shift analysts from energy to other sectors as they fall in and out of favor; consequently, analysts often specialize in an investment strategy (e.g. equity arbitrage, event-driven, macro) rather than a particular sector. Similarly, private equity funds focused exclusively on energy investing are known to hire ex-investment bankers with no energy expertise over seasoned energy experts from operating companies. Despite the enormous size of the investment management field in terms of dollars managed, it is somewhat of a cottage industry, and the majority of jobs are filled through referrals.

Investment analysis starting salaries	
New BA-level:	$50,000 – $60,000
New MBA-level:	$90,000 – $150,000

Consulting

For people with no energy experience, consulting offers a good entry point into the industry. Consulting firms are often willing to hire smart people and train them on the industry content knowledge. BAs with any type of academic concentration are competitive, though some firms may tend to prefer Economics or Engineering students. Given the complex nature of the energy business, PhDs are unusually welcome and explicitly recruited into energy consulting. Most firms start PhD graduates one or two levels below MBAs, given their lack of work experience.

Business consulting firms whose clients are the senior management staff of private companies are called management consulting firms. Working for one of them would get you involved in helping energy company clients answer a wide variety of questions, such as:

- Conduct due diligence for proposed M&A transactions
- Assess the pros and cons of an O&G company investing in capacity to produce and import liquefied natural gas (LNG)
- Advise companies on oil, gas, and electricity price hedging strategies
- Recommend methods for a refinery to cut operating costs
- Discuss alternatives for setting up a joint venture between an investment bank and a utility
- Design a performance management system for a national oil company abroad
- Identify which assets, if any, a utility should divest
- Value power generation assets

Visit Vault at **www.vault.com** for insider company profiles, expert advice, career message boards, expert resume reviews, the Vault Job Board and more.

VAULT CAREER LIBRARY   127

- Calculate the risk management impact of a utility's physical and financial positions

- Evaluate for which oil fields an E&P company should renew leases and pursue development

In contrast to management consultancies, risk consulting firms focus specifically on supporting companies with Value-at-Risk (VaR) management, capital adequacy questions, Sarbanes-Oxley compliance, and establishing enterprise-wide risk management processes. Another niche area within business consulting is litigation support. Litigation support consulting groups (also known as economic consulting firms) provide economic reasoning to support disputes over issues such as price fixing, collusion, or trading improprieties; they may also delve into traditional strategy consulting as well. Many of these firms were originally spun out of Harvard and MIT academic departments, and tend to be staffed with a fair number of PhDs, rather than MBAs.

Hiring for energy practices in consulting firms is generally no different than hiring for other practices – in fact, firms that focus on multiple industries often hire generalists and assign them to a specific industry practice after-the-fact. Like all consultants, energy consultants need to demonstrate excellent problem-solving and client service skills. Interviews for energy-specific consulting jobs will tend to be more quantitative. Because the industry is so asset-intensive, senior executive decisions tend to be heavily based on financial reports and data.

Consulting firms have one of the more rigorous hiring processes among types of companies, using business case interviews to test your ability to quickly assess the probable causes of problems you would likely encounter in client organizations. Anyone attempting to get into consulting is wise to study hard for case questions. Often, the difference between getting an offer and not is more a reflection of how hard you study and how many analysis frameworks you memorize, rather than how intrinsically smart you may be – so don't plan on simply "winging it" in the interview!

You can actually do consulting work in many types of companies. If your work is project-based, as opposed to having a continuous and fixed job responsibility, and you advise people on things, then you are a consultant. In fact, the line between being a consultant and being some type of non-consultant businessperson is quite blurry. Sometimes large corporations have positions for people they actually title "internal consultants." An internal consultant at a large utility company might work on a business strategy project for a few months, and then be asked to help with a new market forecasting initiative, followed by providing valuation support for an environmental compliance decision. Similarly, many services firms do advisory work without referring to themselves as consulting firms per se. A prime example of this is oil services firms, which not only provide outsourced equipment supply services, but also consult to oil companies on exploration tactics and data analysis. Working for an oil services firm could feel very similar to working for a consulting firm, depending on the exact nature of your role.

A Day in the Life: Energy Consulting Associate

9:30 a.m.: Get into the office on a Friday morning and check voicemail and email. You've got a message from the head of the energy practice, asking you to put together a few discussion slides on your practice area's new sales initiative for an internal conference call later in the day. You're happy to be included in this business development work, as it's a great opportunity to get some face time with a few senior partners.

10:00 a.m.: Settling in at your desk with a cup of coffee to wake you up from a week of heavy travel and less-than-optimal sleep, you dig in on incorporating comments from yesterday's conference call into a PowerPoint presentation that's due to the client next week. The partner you are working with is three time zones west of you, so you have an hour and half before you're scheduled to show him your work. The project is heavy on the financial modeling, supporting a $1 billion environmental compliance decision by large utility: should they install scrubbers, or buy SO2 credits?

11:40 a.m.: Conference call with your project manager and partner to review the current version of the draft presentation – the partner only has half an hour, and is running 10 minutes late, so it ends up being a hasty 20-minute call. However, you agree on next steps, and have your marching orders for completing the deck over the weekend.

12:00 p.m.: Today is knowledge-sharing day, and one of your colleagues is presenting findings from a recently-completed M&A valuation project in the conference room. You grab an office-sponsored free sandwich from the kitchen and listen in on the talk.

12:45 p.m.: The receptionist calls you out of the knowledge sharing session to take a client call. The client team leader is upset with the

Visit Vault at **www.vault.com** for insider company profiles, expert advice, career message boards, expert resume reviews, the Vault Job Board and more.

VAULT CAREER LIBRARY   **129**

performance of one of their other vendors on the current project, and asks your firm to help get the task done. You assure him that you will bring it up ASAP with your partner, and figure out a game plan. You leave a voicemail for the partner that you need to talk this afternoon.

1:00 p.m.: Spend some time working on the financial model for the environmental compliance decision project.

3:00 p.m.: Sit on the conference call for which you prepared the discussion slides this morning. It's good that you prepped for it, as the call was quick and efficient. Your group is trying to leverage success on a recent regulatory risk assessment project to create a packaged offering to other energy companies. The partners agree on a split of which prospects each one will call and by when. The upshot of the call for you is that you are to centrally coordinate this new initiative and create all the presentation materials.

4:15 p.m.: Walk downstairs to Starbucks for a quick coffee. You could get some from the office kitchen, but you feel like taking a short break and getting some air.

4:30 p.m.: The partner calls you back in response to your earlier voicemail. You fill her in on the client's issues with the other vendor, and the two of you collaborate on a plan to get the extra work done without blowing your budget. You conference in the vendor team and share the plan with them.

5:00 p.m.: You agreed with your partner that you would fly down to the vendor's offices to manage their work product issue. So you spend a few minutes thinking through the logistics of how best to blend in that trip with your other client travel next week, and then make plane, car, and hotel reservations accordingly. Given the complexity, you decide to just do it yourself, rather than loop in your shared assistant.

5:45 p.m.: Riding the subway home, you read through an energy magazine – industry knowledge is extremely valuable to you as a consultant, and there is always more to learn. After unwinding for a while, you meet up with some friends for dinner.

8:00 p.m.: Spend a couple hours iterating next week's presentation – it'll take several more hours over the weekend to finish it up.

10:00 p.m.: The client calls your cell to follow up on what your resolution is to their problem with the other vendor. You try to conference in the partner, but she's on another call. Fortunately you are able to successfully reassure the client that you are going to fly down and work directly with the vendor firm, and the work will get done. With the client satisfied, you've done a good day's consulting work. Before going to bed, you check your email and voicemail; you reply to one message from a consultant who is updating a financial model you built for another project and had a question about one of your VBA (Visual Basic for Applications) macros.

Consulting starting salaries	
New BA-level:	$40,000 – $55,000
New MBA-level:	$80,000 – $115,000

Business Development

In business schools, business development jobs are the most coveted roles for people not entering the big three services fields (consulting, investment banking, investment management). The hype about "biz dev" exists for good reason: these jobs are strategic and change-oriented. If you thrive on big picture, creative thinking and are a results-oriented and driven person, business development is a compelling functional role to consider in the energy sector.

Business development is a function that all organizations have, but it is not necessarily separated out into its own titular role in all companies. "Biz dev," as it is often called, is concerned with identifying and implementing ways to grow revenue. It involves new product development, market identification, and partnership building. In some firms, the business development function might be housed within a marketing group; however, marketing is typically more downstream and tactically-oriented, focused on post-market optimization of revenue streams rather than creation of new ones. In other firms, the business development function could be housed within a strategic planning group; however, strategic planning is typically more focused on resource allocation and capital investment decisions than on revenue growth strategies:

Visit Vault at **www.vault.com** for insider company profiles, expert advice, career message boards, expert resume reviews, the Vault Job Board and more.

VAULT CAREER LIBRARY **131**

	Business Development	Marketing	Strategic Planning
Typical responsibilities	• New product and service development • Acquisition and partnership strategy • Market assessment	• Pricing, advertising, customer targeting, positioning, competitive intelligence • Product management	• Budgeting across business units • Capital allocation decisions • Financial forecasting
Primary concern	• Where can I find new revenue growth opportunities?	• How can I optimize the sales of my existing products?	• How much do we / should we plan to invest in each business line over the next few years?

In a company that makes physical products – in the energy industry this means the equipment manufacturers and the asset developers-business development is a core strategic function that attracts dynamic, creative people. A business development manager in a startup focuses on how to get the prototype to market. A biz dev team member at a large, established turbine manufacturer is analyzing the potential for entering new geographic markets, and engaged in strategic selling to important bulk-purchase customers and contractor partners. In a power generation company, business developers are likely simply the developers, in charge of identifying sites for new power plants and managing the design, permitting, and economic analysis components leading up to the investment decision.

Professional services firms also often have so-called business development people who are effectively the sales strategists. However, most of the other companies in the energy production value chain – the oil companies, utilities, pipeline operators – tend to not have a business development department. In these large companies that produce BTUs rather than tangible products or professional services, the function of identifying new markets is often housed within strategic planning. Thus, a strategic planning analyst for an oil company would not only coordinate budgeting across business units and generate firmwide financial forecasts, but would also likely analyze revenue potential from new drilling regions, and provide M&A valuation support.

Business development starting salaries	
New BA-level:	$40,000 – $55,000
New MBA-level:	$70,000 – $110,000

Banking

Banking is a very unique type of job in that it is highly transactional in nature. Bankers generally do not get involved in strategic issues related to their industry group, but instead focus on execution of financial deals that can involve billions of dollars. As a result, banking appeals to people who are implementation-oriented, who thrive on pressure, and who have a passion for the market. Theory buffs and people who are fascinated by business strategies and operational issues tend to find banking less exciting.

There is a somewhat blurry distinction between commercial banking and investment banking, particularly since the investment banking industry was further deregulated via the Glass-Steagall Act of 1999. Commercial banks issue loans and lines of credit, but also arrange and underwrite debt capital market transactions just like their investment bank counterparts. In addition to debt capital markets work, investment banks can arrange and underwrite equity market transactions. Commercial banking jobs tend to focus more on issues like credit rating and default risk assessment, whereas investment banking jobs are more about financial strategies and securities pricing. Apart from the debt capital markets overlap area, there is little movement of people back and forth between commercial and investment banking.

Energy work at investment banks is often divided up between a power group and a natural resources group, the latter of which might in turn be bifurcated into oil and gas, and mining. As the amount of investment banking work in the energy sector waxes and wanes, such sub-groups are consolidated and split apart, with ensuing staffing changes.

Energy is a capital-intensive industry, meaning energy companies are constantly in need of funding. Thus, energy bankers have a lot of work in using public equity, debt and private equity markets to raise capital for their clients:

- **Corporate bond issuances:** Companies issue bonds to fund new construction or asset upgrades, refinance existing higher-interest or maturing bonds, and exit from bankruptcy. The investment bank will

Visit Vault at www.vault.com for insider company profiles, expert advice, career message boards, expert resume reviews, the Vault Job Board and more.

VAULT CAREER LIBRARY 133

advise on pricing, and then market or syndicate the bonds to buyers. Transaction values can be $75 million for a small power plant up to more than $300 million for a large, interstate pipeline development, or sometimes billions of dollars for bankruptcy restructurings.

- **Project finance:** Banks use financial engineering to structure complex non-recourse financing packages for new construction. These debt issuances use a project as collateral, rather than the entire corporation that owns the project. Often, these transactions are conducted by a separate group of project finance specialists within the industry group.

- **Equity issuances:** Banks can issue additional stock on behalf of a client company – including initial public offerings (IPOs) of stock. The company then uses the proceeds to recapitalize, reduce its leverage, fund growth, and/or clean up its balance sheet.

- **M&A transactions:** Investment banks do some of their most high-profile work in advising on and brokering asset sales, acquisitions, and company mergers.

For any of these transactions, the investment bank may simply arrange and facilitate it, or may choose to underwrite it (meaning that the bank then agrees to purchase any securities for which it cannot find buyers). In either case, the specific role of the investment banking analyst (pre-MBA level) or associate (post-MBA level) is to build cash projection models in Excel, piece together the information memorandum summarizing the deal specifications to prospective investors, and research potential buyers.

Investment bankers are usually hired as generalists, and only placed into an industry group after coming on board, based primarily on personality fit and secondarily on preference and previous industry experience. For someone who specifically wants to do energy investment banking, the best strategy is to make your preference clearly known during interviews, and then when you start work, lobby the energy group directors with a persuasive case for why you can add value to their group. To get an investment banking job, you don't need a technical degree by any means, but coursework in finance is a must. Generally, the only opportunity to enter the field is right out of college or an MBA program – lateral hires out of other jobs are a purely bull market phenomenon, and rare even then.

Banking starting salaries	
New BA-level:	$45,000 – $55,000
New MBA-level:	$80,000 – $90,000

Strategy and Planning

Strategic planning jobs expose you to an extraordinarily high-level view of a company. Strategic analysts often report up directly to the CEO through their group manager, and are on the inside of some of the most high-impact decisions a company makes. While in some firms, the strategic planning job is more of a glorified secretarial role as regards business unit budget consolidation, in most it is a highly-competitive, high-profile position open to those with a good undergraduate or graduate business education or prior business strategy experience.

Many strategic planning analysts consider themselves internal consultants, with the CEO being their client. They themselves are customers of all of the company's business units, receiving inputs on requested or expected capital expenditures, revenue projections, and business strategies that need to be pulled together into a firm-wide view. Often, they are also the clients of external strategy consulting firms that are brought in to assist and advise on large-scale initiatives. In contrast to the business development role that may also exist in the company, the strategic planning team is focused on investment decisions and resource allocation.

All companies have strategic planning people, whether they are explicitly referred to as such on their business cards or not. Particularly in the case of smaller firms, the strategic planning function may be a part-time responsibility of the corporate finance or business development team. Seeking out these jobs in utilities, oil companies, pipeline operators, and equipment manufacturers requires a lot of due diligence on your part, as each company is highly unique in its structure, and what it may label the people who do the strategic planning work.

Strategic planning roles are at a high enough level that they are not stringent on the industry experience requirement. As a new college graduate, obtaining a strategic planning job gets you great exposure among the senior executives in a company, and allows you to learn a lot about the industry at a macro level. Similarly, coming out of business school into a strategic planning role

Visit Vault at www.vault.com for insider company profiles, expert advice, career message boards, expert resume reviews, the Vault Job Board and more.

VAULT CAREER LIBRARY   135

preserves a fair number of options, in terms of allowing you to later move into finance, business development, or operations management.

Strategy and planning starting salaries	
New BA-level:	$40,000 – $55,000
New MBA-level:	$65,000 – $100,000

Economic and Policy Analysis

Government agencies, think tanks, industry associations, and nonprofit advocacy organizations employ people to analyze economic issues and government policy in the energy sector. (Note: "Nonprofits" are properly, but uncommonly, called "non-for-profit organizations"; they are also referred to as 501(c)3's, referencing the IRS clause which allows qualifying organizations to be tax-exempt.) In addition, economic consulting firms engage in very similar work, often providing outsourced services to government agencies.

One finds a lot of government, political science, and economics undergraduate majors in nonprofit jobs, but also a fair number of other humanities folks as well – the primary requirements for these jobs are a demonstrated passion for the issues, research and writing skills, and a facility with the microeconomics concepts that so fundamentally describe energy sector dynamics. Think tanks, in contrast, are mainly home to PhDs doing academically oriented research work. Government agencies offer a wealth of employment opportunities, from small state energy investment agencies to the massive federal Department of Energy. Most public sector employers offer a variety of positions for new college graduates, experienced economic analysts, and MBAs/PhDs in more senior postings.

Government agencies can be good places to learn a lot about the industry and start off a career. The large federal agencies like the EPA and DOE offer new graduate rotational programs that are a well-respected training ground. People who go into government work may find it difficult to later move into private sector positions without earning another degree like an MBA. However, there is a wealth of interesting positions one can hold over a career within government and nonprofits alone.

In nonprofit organizations, job satisfaction tends to be extremely high. People love the fact that they are impacting government and corporate policy through their daily work, and are often happy to take home part of their paycheck in the form of simply knowing that they are "making a difference." Most nonprofit groups are small, and their energy teams may be just a few people. With so few resources, everyone ends up doing interesting content-oriented work, and it's hard to get shuttled into a "grunt" type of position. But because resources are constrained, nonprofit employees do frequently get burned out by the long hours, and many eventually seek out better-paying jobs after a few years.

Day in the Life: Policy Analyst in a Nonprofit Advocacy Group

8:00 a.m.: Arrive at the office and start the morning with your daily review of the news wires to see what developments are afoot in the energy world: corporate merger rumors, congressional legislation proposals, updates on the latest accounting scandal, announcements by a foreign government about a new infrastructure project or environmental policy.

9:00 a.m.: Your phone rings – it's a Legislative Assistant from a Congressperson's office on "the Hill." She wants to get more information about the press release your group issued last week, which called attention to an energy policy change you and your colleagues found buried in the latest congressional appropriation bill. This LA's question is fairly detailed, so you pass her on to your legal specialist down the hall.

9:15 a.m.: Your main task for today is to call and email a long list of local activist groups, congressional staff, and reporters, spreading your organization's message about the harmfulness of the 11th-hour policy change that ended up in the appropriation bill. You start down the list with the reporters, hoping to get one of them to pick up the story for tomorrow's paper.

12:00 p.m.: The policy director is going to lunch with someone from one of the foundations that funds your organization, and asks you to come along to discuss the initiatives you've been working on lately. While the energy policy bill is your latest overwhelming concern, this foundation wants to hear about work on the topics it earmarked its funds for: a recent campaign about municipal waste

Visit Vault at **www.vault.com** for insider company profiles, expert advice, career message boards, expert resume reviews, the Vault Job Board and more.

VAULT CAREER LIBRARY   **137**

incineration emission standards, and an ongoing study about nuclear waste disposal alternatives (of which you are to be the primary author).

4:00 p.m.: Fielding and placing calls related to last week's press release has taken up your whole afternoon. Now you carve out a few hours to work on background research for that nuclear waste study – you have a hefty reading list to get through, including a stack of company annual reports, magazine articles pulled from an online research database, and existing studies by other research organizations.

8:00 p.m.: After a long day, you walk down the street to the neighborhood bar and have a few beers with your friends who work at other nonprofits. Your college roommate had emailed an invitation to a cocktail hour at a posh new lounge, but you declined in favor of $1 drafts. Sometimes it bothers you to not be able to keep up with your banker buddies' lifestyles, but then again, you find your chosen career very fulfilling – while they chat over martinis about the prospect for a next promotion, you will enjoy an animated intellectual debate about the virtues of offshore windpower development.

Economic and policy analysis starting salaries	
New BA-level:	$20,000 – $35,000
New MBA-level:	$35,000 – $80,000

APPENDIX

Resources

Below is a list of resources that you may find useful as you continue building your energy knowledge and continue exploring careers in energy. Please note that Vault does not specifically endorse or have a business relationship with any of these (besides www.vault.com).

Periodicals

Electricity Journal (An informative monthly with practical articles written by a mixture of consultants, industry executives, and academics. www.electricity-online.com)

The Economist (A favorite of many businesspeople, this widely read weekly includes timely articles covering energy sector issues and their impact on international economies and politics. www.economist.com)

Houston Chronicle (This newspaper is known to have excellent ongoing coverage of the oil and gas and electricity sectors. Of particular note was their detailed yet highly accessible documentation of the Enron scandal. www.chron.com)

Public Utilities Fortnightly (A comprehensive publication covering the electricity and natural gas business, technology, and regulation. www.pur.com/puf.cfm)

Energy Risk (Formerly known as EPRM, this magazine focuses on the risk management side of electricity and oil & gas. www.eprm.com)

Project Finance Magazine (Covers project finance issues related to oil, gas, and power projects, as well as other large infrastructure projects. www.projectfinancemagazine.com)

Books

Energy: Physical, Environmental, and Social Impact. Gordon Aubrecht, 2005. (A comprehensive textbook on the principles of energy production, cost, storage, and conservation)

Pipe Dreams: Greed, Ego, and the Death of Enron. Robert Bryce, 2002. (One of the more detailed post-mortems on the 2001 demise of the country's largest energy trading firm)

Modeling Prices in Competitive Electricity Markets (The Wiley Finance Series). Derek Bunn, ed., 2004. (Detailed how-to on electricity price forecasting methods)

Energy and Power Risk Management: New Developments in Modeling, Pricing and Hedging. Alexander Eydeland and Krzysztof Wolyniec, 2002. (Definitive guide to energy asset valuation and risk management strategies)

Wind Energy Comes of Age. Paul Gipe, 1995. (The definitive textbook on wind energy engineering, operation, cost, and environmental impacts.)

Tomorrow's Energy: Hydrogen, Fuel Cells, and the Prospects for a Cleaner Planet. Peter Hoffman, 2002. (A thorough exposition of the prospects for using hydrogen as an energy storage device)

Energy Management Handbook. Wayne Turner and Warren Heffington, 2004. (Detailed reference information on operational management of commercial and industrial energy systems.)

Power Generation, Operation, and Control. Allen Wood and Bruce Wollenberg, 1996. (Engineering textbook covering the mechanics, costs, and operational management issues of electric power plants)

The Petroleum Industry: A Nontechnical Guide. Charles Conaway, 1999. (Comprehensive introduction to petroleum geology, drilling techniques, production and distribution methods)

Online reference materials

Policy

Federal Energy Regulatory Commission (FERC): The FERC is an independent agency that regulates the interstate transmission of natural gas, oil, and electricity, as well as projects concerning natural gas and hydropower projects. Check out their eLibrary and Students' Corner sections. (www.ferc.gov)

Environmental Protection Agency (EPA): Visit the EPA's website to see the latest progress regarding emission controls legislation. (www.epa.gov)

World Energy Council: Home page for the largest global energy policy organization, with members in over 90 countries. (www.worldenergy.org)

Rocky Mountain Institute (RMI): RMI's website has many resources on sustainable energy policy. (www.rmi.org)

Pew Climate Center: Pew's website provides ample background and current information on climate change and related policy progress. (www.pewclimate.org)

Power/Electricity

Edison Electric Institute: Major trade association focusing on utilities, with some outstanding downloadable primers on the industry geared towards the public at large. (www.eei.org)

Energy Information Administration (EIA): The statistical agency of the U.S. Department of Energy whose website houses a wealth of energy data, forecasts, and analyses. (www.eia.doe.gov)

Energy Central: Contains comprehensive daily news focused on the global power industry. (www.energycentral.com)

Oil and Gas

Oil and Gas International: A subscription-based content website with coverage of E&P worldwide. (www.oilandgasinternational.com)

Herold: A leading subscription-based newsletter service focused on oil and gas. (www.herold.com)

Rigzone: An online portal for oil and gas industry information (www.rigzone.com)

Society of Petroleum Engineers (SPE): Knowledge base and event listing for E&P. (www.spe.org)

American Association of Petroleum Geologists (AAPG): Website containing the latest news and events from an organization whose mission is to advance the science and technology of petroleum geology. (www.aapg.org)

General

Eye for Energy: A solid online resource with news and analysis for E&P, transportation, refining, storage & distribution, trading & marketing, renewables/green power, and utilities generation, transmission and distribution. (www.eyeforenergy.com)

World Energy News: Comprehensive global energy news coverage. Unlike other websites, has many useful links on international energy. (www.worldenergynews.com)

Visit Vault at **www.vault.com** for insider company profiles, expert advice, career message boards, expert resume reviews, the Vault Job Board and more.

V\ULT CAREER LIBRARY   **143**

Job listing Sources

You probably know by now that there are literally thousands of online job sites, and they seem to appear and disappear with startling frequency. Here are a few of the most popular and stable ones.

Energy job sites

- www.rigzone.com/jobs (Oil and gas industry)
- www.energycentraljobs.com
- www.energyjobsnetwork.com (Covers Europe as well as U.S.)
- www.thinkenergygroup.com (Electricity industry)
- www.globalenergyjobs.com

General job sites (These allow you to search or browse by industry but do not have a specific energy focus.)

- www.hotjobs.com
- www.monster.com
- www.careerbuilder.com
- www.dice.com (Technology and engineering-oriented jobs)
- www.employmentguide.com
- www.craigslist.com
- www.flipdog.com
- www.jobs.com
- www.jobsinthemoney.com (Finance-oriented jobs)
- www.glocap.com (The best source for hedge fund, private equity, and investment banking openings)
- www.analyticrecruiting.com (Finance, risk management, marketing science, operations research, and quant jobs)
- www.vault.com (Includes a job board with postings about individual companies)

Major Employers

Below is a list of some major employers in each company category. This list focuses on U.S.-based companies; however, given the international nature of the energy business, many of them have international offices.

Oil Companies

The majors

Integrated oil and gas companies

- BP (London, www.bp.com)
- ExxonMobil (Irving TX, www.exxonmobil.com)
- Shell (The Hague, www.shell.com)
- ChevronTexaco (San Ramon, www.chevrontexaco.com)
- ConocoPhillips (Houston, www.conocophillips.com)

Dedicated purely to exploration and production

- Gulf (Chelsea MA, www.gulfoil.com)
- Occidental (Dallas, www.oxy.com)
- Unocal (El Segundo CA, www.unocal.com)

The independents – mostly located in the U.S., where inland exploration is prevalent. Offshore E&P is too expensive for all but the largest companies.

- Burlington Resources (Houston, www.br-inc.com)
- Devon (Oklahoma City, www.devonenergy.com)
- Apache (Houston, www.apachecorp.com)
- Anadarko (Houston, www.anadarko.com)

Refiners – these firms focus on midstream and downstream oil & gas activities, i.e. refining and marketing.

- Valero (San Antonio, www.valero.com)
- CITGO (Houston, www.citgo.com)
- Sunoco (Philadelphia, www.sunocoinc.com)

Visit Vault at **www.vault.com** for insider company profiles, expert advice, career message boards, expert resume reviews, the Vault Job Board and more.

VAULT CAREER LIBRARY 145

Mining

- Rio Tinto (London, www.riotinto.com)
- BHP Billiton (Melbourne, www.bhpbilliton.com)

Oil Services Companies

Note that these are sometimes loosely referred to as "energy services," but that has another specific meaning in the power sector.

- Halliburton (Houston, www.halliburton.com)
- Schlumberger (New York, www.slb.com)
- Atwood Oceanics (Houston, www.atwd.com)
- Noble (Sugar Land TX, www.noblecorp.com)
- Nabors Industries (Barbados, www.nabors.com)
- TransOcean (Houston, www.deepwater.com)

Pipeline Operators

These companies are common carriers of petroleum and/or natural gas products, and often also operate refineries and oil terminals.

- Enbridge (Calgary, www.enbridge.com)
- Buckeye Partners (Emmaus PA, www.buckeye.com)
- Valero (San Antonio, www.valero.com)
- Kinder Morgan (Houston, www.kindermorgan.com)
- TransCanada (Calgary, www.transcanada.com)
- Williams (Tulsa, www.williams.com)

Utilities

- Exelon (Chicago and Kennett Square PA, www.exeloncorp.com)
- PSE&G (Newark NJ, www.pseg.com)
- AEP (Columbus, www.aep.com)
- FirstEnergy (Akron OH, www.firstenergycorp.com)
- Edison International (Rosemead CA, www.edison.com)
- Dominion (Richmond VA, www.dom.com)

- Southern Company (Atlanta, www.southernco.com)
- PG&E (San Francisco, www.pge.com)

Transmission Grid Operators

- Independent System Operators (ISO) and Regional Transmission Operators (RTO)
- ERCOT (Austin TX, www.ercot.com)
- California ISO (Folsom CA, www.caiso.com)
- ISO New England (Holyoke MA, www.iso-ne.com)
- NY ISO (Schenectady NY, www.nyiso.com)
- PJM (Norristown PA, www.pjm.com)
- Midwest ISO (Carmel IN, www.midwestiso.org)

Energy Equipment Manufacturers

Turbines

- ABB (Zurich, www.abb.com)
- GE (Fairfield CT, www.ge.com)
- Westinghouse (Monroeville PA, www.westinghouse.com)

Fuel cells

- Ballard Power Systems (British Columbia, www.ballard.com)
- General Hydrogen (British Columbia, www.generalhydrogen.com)
- Stuart Energy (Ontario, www.stuartenergy.com)
- UTC Fuel Cells (South Windsor CT, www.utcfuelcells.com)
- Plug Power (Latham NY, www.plugpower.com)
- Proton Energy Systems (Wallingford CT, www.protonenergy.com)
- Shell Hydrogen (Amsterdam, www.shellhydrogen.com)
- PowerZyme (Princeton, www.powerzyme.com)

Visit Vault at **www.vault.com** for insider company profiles, expert advice,
career message boards, expert resume reviews, the Vault Job Board and more.

VAULT CAREER LIBRARY   **147**

Pollution control technologies

- Green Fuel Technologies (Cambridge MA, www.greenfuelsonline.com)
- Catalytica Energy Systems (Mountain View CA, www.catalyticaenergy.com)
- H2Gen (Alexandria VA, www.h2gen.com)

Oil and gas production equipment

Note that many companies in this space also provide oil services, and you may find them categorized as such in your research.

- Cooper Cameron Corporation (Houston, www.coopercameron.com)
- Oil States International (Houston, www.oilstates.com)

Information management

- Itron (Spokane, www.itron.com)
- Softricity (Boston, www.softricity.com)
- SmartSynch (Jackson MS, www.smartsynch.com)

Investment Funds with Significant Energy Exposure

Mutual funds

- Fidelity Utilities Fund
- Franklin Templeton Utilities Fund
- Oppenheimer Real Asset Fund (Energy and Natural Resources)
- T. Rowe Price New Era Fund (Energy and Natural Resources)
- Vanguard Energy Investment Fund (Energy and Natural Resources)

Hedge funds

Hedge funds change their sector weightings constantly and do not report their investment philosophies publicly, so it is very difficult to identify which ones have lots of energy-related activity.

- Vega Asset Management – largest hedge fund in world; 2% of its capital, or $300m, is invested in energy (Madrid)

• Elevate Capital Management (Short Hills, NJ)

• Houston Energy Partners (Houston, www.smhgroup.com)

• Eiger Energy Trading (Geneva, www.eigertrading.com)

Private equity firms

• Nth Power (San Francisco, www.nthpower.com)

• Enertech Capital (Wayne PA, www.enertechcapital.com)

• Ridgewood Capital (Ridgewood NJ, www.ridgewoodcapital.com)

• Rockport Capital (Boston, www.rockportcap.com)

• Altira Group (Denver, www.altiragroup.com)

• FA Technology Ventures
  (Albany NY and Boston, www.fatechventures.com)

• Advent International (Boston, www.adventinternational.com)

• Siemens Venture Capital
  (Munich and Boston, www.siemensventurecapital.com)

• DTE Energy Ventures (Detroit, www.dteenergyventures.com)

• GFI Ventures (Los Angeles, www.gfienergy.com)

Banks with Large Energy Practices

Investment banks

Bulge bracket

• Lehman Brothers (New York, www.lehman.com)

• Goldman Sachs (New York, www.gs.com)

• Merrill Lynch (New York, www.ml.com)

• Morgan Stanley (New York, www.morganstanley.com)

Specialty boutiques

• Simmons & Co. (Houston, www.simmonsco-intl.com)

• Waterous & Co. (Calgary, www.waterous.com)

Visit Vault at **www.vault.com** for insider company profiles, expert advice,
career message boards, expert resume reviews, the Vault Job Board and more.

VAULT CAREER LIBRARY   **149**

Commercial banks

- ABN AMRO (Amsterdam, www.abnamro.com)
- Citibank (New York, www.citigroup.com)

Consulting Firms

Buyers of consulting services are trending more and more towards firms with deep industry expertise versus general strategy prowess. As a result, a list of top providers of energy consulting services will look slightly different than a generalized "top ten" list.

Management consulting

Some of the major strategy consulting brands don't do significant energy work (e.g. Bain, BCG).

- McKinsey (New York, www.mckinsey.com)
- Booz Allen Hamilton (New York, www.bah.com)
- Strategic Decisions Group (Palo Alto CA, www.sdg.com)
- Navigant Consulting (Chicago, www.navigantconsulting.com)
- Accenture (Chicago, www.accenture.com)
- Global Energy Decisions (Boulder CA, www.globalenergy.com)

Risk consulting

- Marsh (of the consulting giant Marsh & McLellan) (New York, www.marsh.com)
- PricewaterhouseCoopers (New York, www.pwcglobal.com)
- Aon Consulting (Chicago, www.aon.com)

Economic consulting

- Charles River Associates (Boston, www.crai.com)
- NERA (Boston, www.nera.com)
- Lexecon (Chicago, www.lexecon.com)
- London Economics Consulting Group (Boston, www.londoneconomics.com)
- ICF Consulting Fairfax VA, www.icfconsulting.com)

• Industrial Economics (Cambridge MA, www.indecon.com)

Nonprofits

Each organization only employs a handful of people who deal with content issues, while a large fraction of the staff takes care of fundraising and administration.

- Rocky Mountain Institute (Snowmass CO, www.rmi.org)
- Public Citizen (Washington DC, www.citizen.org)
- Natural Resources Defense Council (NRDC) (New York, www.nrdc.org)
- Environmental Defense Fund (EDF) (New York, www.edf.org)
- World Resources Institute (Washington DC, www.wri.org)
- American Council for an Energy Efficient Economy (Washington DC, www.aceee.org)
- Alliance to Save Energy (Washington DC, www.ase.org)
- Union of Concerned Scientists (Cambridge MA, www.ucsusa.org)

Government Agencies

Federal

- Federal Energy Regulatory Commission (FERC). 1,200 employees, based in Washington DC (www.ferc.gov)
- North American Energy Reliability Council (NERC). 50 employees, based in Princeton, NJ (www.nerc.com). (Note that there are ten Regional Reliability Councils as well: ECAR, ERCOT, FRCC, MAAC, MAIN, MRO, NPCC, SERC, SPP, and WECC)
- Nuclear Regulatory Commission (NRC). 2,500 employees at Rockville, MD HQ, plus regional offices (www.nrc.gov)
- Environmental Protection Agency (EPA). 5,000 employees in the Office of Air, which deals primarily with the energy sector, given that it is the primary air polluter; HQ in Washington, DC, plus regional offices (www.epa.gov)
- Department of Energy (DOE). 1,000 employees, based in Washington DC, plus field offices and labs (www.doe.gov)

Visit Vault at **www.vault.com** for insider company profiles, expert advice, career message boards, expert resume reviews, the Vault Job Board and more.

VAULT CAREER LIBRARY   **151**

State

- State departments of environmental protection (DEPs). 4,000 employees across the country in the air divisions)

- State public utility commissions (PUC) and public service commissions (PSC). 20,000 employees across the country.

- State development authorities. Some states have government authorities (separate from the state DOE or DEP) to promote efficiency and renewable energy investment. They invest money from energy bill surcharges on behalf of the public interest. These organizations are little-known gems for obtaining quasi-VC experience. 5,000 employees across the country.

- New York State Energy Research and Development Authority (Albany NY, www.nyserda.org)

- California Energy Commission (Sacramento CA, www.energy.ca.gov)

- Energy Office of Michigan (Lansing MI, www.michigan.gov/cis)

- Massachusetts Renewable Energy Trust (Westborough MA, www.mtpc.org)

- Energy Trust of Oregon (Portland OR, www.energytrust.org)

- Ohio Energy Loan Fund (Columbus, www.odod.state.oh.us/cdd/oee/energy_loan_fund.htm)

Energy Services Firms

- Ameresco (Framingham MA, www.ameresco.com)

- Noresco (Westborough MA, www.noresco.com)

- Siemens Building Technologies (Munich, www.siemens.com)

- Chevron Energy Solutions (San Francisco, www.chevronenergy.com)

- Onsite Energy Corporation (Carlsbad CA, www.onsiteenergy.com)

Companies with Significant Asset Development Activities

General power plant development

- Florida Power and Light (FPL) (Juno Beach FL, www.fpl.com)

- PPM Energy (Portland OR, www.ppmenergy.com)

- Calpine (Houston, www.calpine.com)
- AES (Arlington, VA, www.aes.com)
- GE (Stamford CT, www.gepower.com)

Windpower development

- Airtricity (Dublin, www.airtricity.com)
- PPM Atlantic Renewable (Portland OR, www.ppmenergy.com)
- EnXco (North Palm Springs CA, www.enxco.com)
- Energy Northwest (Richland WA, www.energy-northwest.com)
- Midwest Renewable Energy Corporation (Joice IA, www.midwest-renewable.com)
- TradeWind Energy (Lenexa KS, www.kansaswindpower.com)
- Navitas Energy (Minneapolis, www.windpower.com)
- Cielo Wind Power (Austin TX, www.cielowind.com)
- Northern Power Systems (Waitsfield VT, www.northernpower.com)
- Zilkha Renewable Energy (Houston TX, www.zilkha.com)

Nuclear development

- Exelon Generation (Chicago and Kennett Square PA, www.exeloncorp.com)
- Dominion Energy (Richmond VA, www.dom.com)
- Entergy Nuclear (New Orleans, www.entergy-nuclear.com)

Selected Companies Energy Trading and/or Energy Marketing Activities

Companies

- Constellation (Baltimore, www.constellationenergy.com) – the largest energy trader in the U.S.
- UBS (Zurich and New York, www.ubs.com)
- Morgan Stanley (New York, www.morganstanley.com)
- Reliant (Houston, www.reliant.com)
- AEP (Columbus, www.aep.com)

Visit Vault at **www.vault.com** for insider company profiles, expert advice, career message boards, expert resume reviews, the Vault Job Board and more.

VAULT CAREER LIBRARY   **153**

- Duke (Charlotte NC, www.duke-energy.com)
- Mirant (Atlanta, www.mirant.com)
- Cinergy (Cincinnati, www.cinergy.com)
- Dynegy (Houston, www.dynegy.com)
- Exelon (Chicago and Kennett Square PA, www.exeloncorp.com)
- Coral Energy (Houston, www.coral-energy.com)

Commodities exchanges

- Intercontinental Exchange (ICE) (Atlanta, www.intcx.com)
- Chicago Mercantile Exchange (CME) (Chicago, www.cme.com)
- New York Mercantile Exchange (NYMEX) (New York. www.nymex.com)
- Chicago Climate Exchange (CCX) (Chicago, www.chicagoclimatex.com)

About the Author

Laura Walker Chung is a pioneer in renewable energy, having conceived and developed the first wind power generation facility for Pacific Gas and Electric's independent power arm. A rare female veteran of the energy sector, she has excelled in energy-related roles in project development, price forecasting, management and economics consulting, and corporate finance. Currently, she is a project leader at a management consulting firm. In her spare time she is a freelance writer, abstract painter, and interior designer. She lives in Cambridge, Massachusetts with her doting husband Eric and equally doting puppy Brian. Laura graduated with highest honors from Dartmouth College and earned her MBA in finance and economics from the University of Chicago Graduate School of Business.

Use the Internet's
MOST TARGETED
job search tools.

Vault Job Board

Target your search by industry, function, and experience level, and find the job openings that you want.

VaultMatch Resume Database

Vault takes match-making to the next level: post your resume and customize your search by industry, function, experience and more. We'll match job listings with your interests and criteria and e-mail them directly to your inbox.

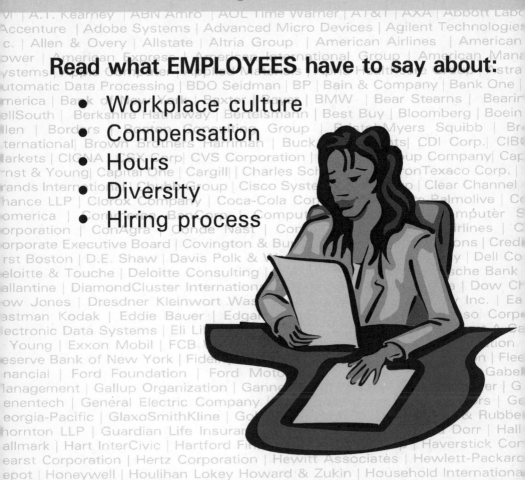